Raising Goats for Beginners

The Complete Guide to Raising and
Caring for Dairy and Meat Goats

David Oliver

Table of Contents

Introduction

I grew up on a small farm with my father, the two of us raising goats in the sunny countryside. Over time, I came to appreciate our livestock for their productive capabilities and thought it was odd just how little we see goat-based products in the mainstream world. After all, it was only natural that I'd think drinking goat's milk and eating their meat wasn't unusual, having grown up on it.

My father and I had a wonderful location to work from, too; the farm was stationed in a rural area just outside a small town where we could easily run in and get supplies both for ourselves and our livestock.

Now that I'm older and still care for goats as a hobby, I wanted to share everything I've learned from raising them so that others can experience the joy I feel watching the little creatures I raised become playful, head-butting, affectionate animals. Hobby livestock owners that own chickens may be well-suited for this new path in particular since those that find joy in watching their chickens play will most definitely love the antics of the goat.

Before I start putting what I've learned to paper, I want to make something very clear. I have not, nor have I ever commercially owned livestock for the sake of major profit. The information you read in this book may not be compatible with becoming a large-scale rancher with hundreds of goats to call your own.

If I were an established person in the world of livestock trade, I'm sure the advice I could give and write about would be a lot more concise. However, with my humble origins, everything that I have learned has been home-grown and "grass fed." If you find better information from professional goat farmers, then don't be afraid to use it. See this book as a springboard to start your new hobby of raising goats and discovering what breed, temperament, and production needs are best for you. Most of all, though, my hope is that you will have fun doing it!

Having a new goat is not unlike having a new puppy. There are plenty of pros and cons, and in the end, you can either have a fun, friendly farm animal or a nasty hip-height torpedo on legs aiming right for you. Remember that with every step into a new hobby, there will be a margin for error. If you feel at any point that you have made a wrong choice or a mistake, don't panic; correct the issue as calmly as you can. Most goats will forgive you, and those that don't work just as well on the dinner table.

CHAPTER ONE:
Breeds of Goats

Before running out the door to meet your new farm animal, you should first figure out exactly what you want to get out of owning a goat. Do you want one that will produce lots of milk for you to use in food or to make soaps and lotions with? If that doesn't appeal to you, would you like to raise goats for the sake of harvesting your own meat from them? Whichever types of goats you prefer to raise, there is a sense of comfort that comes from raising an animal ethically. In this case, I specifically refer to treating the animals you plan to raise with respect before, during, and after you receive from it whatever you hope to gain in its production.

In the modern American lifestyle, much of our meat is created using unethical means. Cattle are being fed hundreds of pounds of corn daily, and chickens have been bred to mature so quickly that their heavy bodies can't be supported by their own legs (Kenner 2009).

I mention this not to shame anyone for participating by buying meat at the grocery store; after all, I'm guilty of this, too. Rather, I bring this up because if you're considering raising goats for meat over dairy purposes, it could be seen as a more eco-friendly way to produce meals.

On that note, a memory is recalled to me that I want to share here. I once had the pleasure of speaking with a pescatarian who told me his reason for eating only fish as his meat choice was due to the ethical means with which fish are farmed. Out of curiosity, I asked him if he would ever eat meat that was hunted or self-farmed for the sake of food. It gave him pause. In the end, he agreed that he would partake of eating meat only if it came from such a source, and he hadn't considered that to be a viable option to fit the parameters of his diet choice.

The answer he gave for being a pescatarian was one of the most interesting reasons I'd ever heard from anyone choosing to eat a specific diet. There is an amount of respect I have for those who choose to ethically harvest their meat or dairy products. In this case, I have to wonder: would some vegans mind if they had dairy from happy, healthy goats they own, too? That's something I'm left to ponder on and for you, the reader, to decide.

For whichever breed you select, you may discover surprising things about yourself as you raise them over time.

BREEDS OF DAIRY GOATS

There are dozens of reasons why one would choose to raise dairy goats: allergies to cow's milk, a taste for it over the former, a desire to make crafts like soap and lotions, just to name a few. Goat's milk is rich in protein, making it an overall healthier choice for consumption.

The industry for goat's milk has only really started picking up since the early 1900s, which means there's still plenty of potential for growth in the market. Regardless of your reason, there's no question that these goats are a great investment for anyone who is fond of milk but doesn't have the desire to own or purchase dairy cows.

For more information as to how to milk goats or how to select a goat from the market when being mindful of their quantity of production, please refer to chapter eight.

Alpine

First imported to the US in 1922, Alpine dairy goats are most likely what you think of when the words "dairy goat" come to mind. They are widely considered to be friendly, graceful, and curious. As one of the largest dairy breeds, they can weigh between 130 and 160 pounds. These lovely goats are short-haired, come in many different colors, and may produce up to three gallons of milk daily. Alpines, which can be either French or American, are an excellent goat to start with due to their easy-going nature (Knapp 2021).

Guernsey

"Golden Guernsey Goats," as they're often called, are astoundingly new to the market. Having been discovered first living feral on Guernsey Island in the English Channel, these beauties may have gone extinct if not for the helping hand of Miriam Milbourne, a woman credited with saving the species in 1924.

Guernsey goats were first introduced to the U.S. in the late 1990s and are still considered to be very rare. True to their nickname, Guernsey goats come in all shades of gold with long hair on their bellies to give them a "skirt," making them resemble highland cattle. Their ears are bonnet-shaped, and they may weigh anywhere between 120 and 150 pounds.

Guernseys are best for those who are patient since the goats have a naturally feral nature (being so new to domestication), but they will in time become more docile. Another great reason to own a Guernsey is if you don't have many acres to dedicate to livestock, as these goats tend to be excellent foragers. As Zenzi Knapp of Morning Chores says, "If methods such as rotational grazing and no-grain livestock are important to you, Golden Guernsey goats are your friend" (2021).

Take note that at their peak, Guernsey dairy goats only produce just under 1.5 gallons of milk daily. To make up for it, they have a high butterfat content, making their milk great for creating homemade dairy products (Knapp 2021).

LaMancha

When you see a LaMancha goat, your first thought might just be, "Whoa, that goat doesn't have any ears!" and you'd be somewhat correct! These goats tend to have very little to show for the sides of their head, having either "gopher" or "elf" style ears that are very small when compared to their wide head.

Bred by Eula Fay in California, it wasn't until 1958 that LaMancha goats became an officially recognized breed. Being quite hardy, they're able to adapt to most climates with the exception that they are short haired, and thus would need protection from harsh winter chills. Weighing between 130 and 155 pounds, they may be potentially used for meat. Their dairy production is up to three-fourths of a gallon daily during their peak season.

What they lack in ears, they make up for in personality. LaMancha goats are extremely friendly and love to socialize, making them perfect for an owner who will spend lots of time petting them. There is a con to this, however, as due to their love of socializing, LaMancha goats will do their best to get out of their pen to find someone to give them affection, so be mindful to put them in a pasture with a strong fence. (Greer 2021)

Nigerian Dwarf

While most goats are on average around 30 inches tall, smaller breeds, such as the Nigerian dwarf, only grow between 16-23 inches in height. These cute, typically solid-colored goats are in high-demand for their quaint nature and are thus often very expensive in the market—though their size matches their dairy output at approximately one to three quarts daily. However, they are said to produce the sweetest and creamiest milk of any breed, and are often content with grazing from the pasture (or "mowing" your yard).

While all goats are natural escape artists, Nigerian Dwarfs are particularly slippery due to their size, meaning special fencing is in order. It is also recommended that if you choose to start with these goats, you only pen them with other miniature breeds. If placed into a pen larger breeds, you risk the goats attempting to mate; a situation in which the miniature goat will more than likely get hurt. They are very playful and better off with someone who already has a little experience with raising goats (Poindexter, 2021).

Nubian
You've heard of lop-eared rabbits, now get ready for lop-eared goats! These gorgeous creatures have long, drooping ears on either side of their head and are quite hardy for all climates; making them possibly one of the best all-rounded goats one can own. The Nubian goat may produce up to one gallon of milk a day with a high butterfat content. This often makes them the best choice if an owner seeks to make their own products such as cheese or ice cream.

Nubians can weigh between 130 and 160 pounds and are one of the largest dairy goats on the market. They're the most vocal of dairy breeds—making them a less desirable breed if you live somewhere with neighbors who will complain about the noise level—but they make up for it by being very affectionate. Nubians are perfect for any owner who has an established farm with plenty of land for goats to vocalize as much as they like (Knapp 2021).

Oberhasli
Sometimes called the "Swiss Alpine" goat, Oberhasli goats were recognized as an official goat breed in 1979. This breed is undoubtedly the best for milk production, and is the usual choice for commercial milk farmers when it comes to goat milk. Their colors are very charming, being medium-haired with variations of red and black. Weighing between 120 and 150 pounds, Oberhasli goats are known to be friendly, gentle, quiet, and very competitive with others in their herd (Allen 2021).

Saanen
The Saanen breed is particularly interesting due to its multi-purpose nature. While it may be used for dairy and meat, it is the unsung hero of pack goats.

Pack goats are ideal for hikers who wish to unload their burdens on animal companions while they travel, hence being akin to a "pack mule." Saneens can hold up to 25% of their body weight, making them ideal for the nomadic, traveling farmer once they've been trained to pack properly. Or, if you're more of a homebody, they're fantastic for helping you bring your supplies around your yard—almost like a more agile, friendly wagon.

These goats are very popular, and for good reason. Their typically white or cream colored fur makes them gorgeous to look at, and they are the absolute best producers for dairy with up to three gallons daily. This is because the Saanen goat has been bred for the absolute best in production since the early 1900s.

They are quieter than most breeds, and have a very calm temperament which makes them a great choice for introducing children to goats. Perhaps the most notable benefit to Saanens, however, is their hardiness. Despite their fragile appearance, these goats are quite hardy creatures and one of the least likely breeds to get sick. The biggest downside is that these goats are notorious for digging—meaning any enclosure you put them in should have a partially buried fence to prevent this means of escape (Pieper 2021).

Sable
Another breed of goat, known as the Sable, shares all characteristics with the Saanen with the exception of color. Saanen goats can only be white or light cream in color, while Sables can be any other color except the latter.

Toggenburg
Named after its region of origin in Switzerland, the Toggenburg is one of the oldest, most recognizable dairy goats on the market. Although it's a good producer, making three-fourths of a gallon daily, it was once unpopular as a sole source of milk due to its flavor and low fat or protein content. Due to selective breeding, however, it is now one of the best producers of milk with a high fat content at 3.7%.

As far as goats go, Toggenburgs are one of the most straight-forward you can find in personality. They are described as "all business" and probably the least playful of any dairy goat breed. This makes them ideal if you want a larger herd of goats and can't handle their springy, bouncy personality while trying to maneuver them in large quantities for milking. Toggenburgs possess white hooves with colors ranging from a rich brown to a light chocolate (Kaminski-Killiany 2017).

BREEDS OF MEAT GOATS

The American Goat Association states, "Goat meat is leaner than poultry and other red meats, low in fat and cholesterol and is a source of conjugated linoleic acid." Although goat meat is one of the most commonly consumed meats in the world, it has historically been more common in the Middle East, Europe, and Asia than in the USA. Due to the melting-pot culture of America, however, the USA now imports 50% of all goat meat eaten on the planet. Equivocal in calories to poultry but with less fat and saturated fat—with the lowest cholesterol, highest protein and highest iron contents of all meat—it's no wonder it's being hailed as a food for anyone looking to guard their health.

From a goat, you are able to craft cuts of meat such as legs, ribs, shoulder, loin roast, breast, stew meat, riblets, and shanks. In addition, you can easily create ground meat to substitute for ground beef in any recipe it appears in.

Goat meat can be called two things. The first is Cabrito, meat from a young goat between four and eight weeks old. Chevon is the name of goat meat from a goat that is 48-60 lbs and between six to nine months old (AGF 2021).

Depending on the breed of the goat, the general rule is that you want the goat you butcher (aside from the young ones just mentioned) to weigh between 80-100 lbs and be eight to ten months old (McCune 2021). It is possible to eat an older goat, with the caveat that it will not be as flavorful or tender. Larry Jacoby of Shepherd Song Farm in an interview with Heavy Table spoke on this fact, stating: "Goat meat is very delicate, really, unless you eat an old animal. That old buck goat is going to take a lot of spices and garlic.... Actually, people don't like to hear it, but bad goat is like good venison." (Pagani 2012)

Spanish

Requiring meat for the journey, Spanish explorers brought along with them goats to North America in the 1500s. Finding other meat sources or losing many of their herd, most of the goats were lost or allowed to go feral in the wild—thus making the Spanish goat an invasive species that has been around for several centuries in North America.

Since then, they have become very hardy and adaptable to otherwise unfriendly environments for livestock. They are also considered disease- and parasite-resistant. Little has been taken to care for their maintenance as they've almost primarily been used to develop other breeds in animal husbandry.

Because of the amount of breeding from these goats, there is a good amount of diversity in the genetics. Due to this, there is variety in their appearance, and Spanish goats come in all colors. ranging from long-haired to short-haired with a cashmere coat. It is easy to come across cross-bred Spanish goats, but a purebred one will have horns swooping up and outward with horizontal ears that are close to the head and face. They can reach up to 20 inches in height and 250 pounds for bucks, while does may weigh up to 150 pounds.

The breed is comparatively flighty beside other types of meat goat breeds and is notoriously nervous of humans. Their feral blood may make it difficult to work with them, so they are not recommended for most beginners (Pagani 2021).

Boer

Along with the Kiko, Boer is undoubtedly one of the most popular meat goats on the market. It is best known for its meat-to-bone conversion and quick growth, meaning you can get it on the market or harvest its meat faster than most other breeds. It is the most common breed used in the goat-meat industry for mass production, with bucks weighing up to 340 pounds and does up to 230 pounds.

A traditional Boer is usually white with a brown head, but they may come in black, white, and brown, as well. The ears are long-drooping, and the body is very bulky in its girth. The average amount of meat gained from these creatures ranges between 40-50 pounds at approximately eight months of age.

Though hardy to many different environments, they are known to have issues with parasites more so than any other breed. Have a deworming action plan before starting to raise them and be cautious when purchasing, as you may end up with a sickly goat.

Boers are gentle giants; they enjoy attention and are overall quite docile (Knapp 2021). They're considered to be one of the best pack goats, so if you're going on a hiking trip and want a goat companion with you, you can't go wrong with a Boer by your side (Pieper 2021). (Just don't get too attached, or you might not end up wanting to send them to the butcher!)

Kiko

Bred by carefully crossing lineages of dairy goats with feral goats, the Kiko is second in popularity only to the Boer. It is a very maternal breed, being very good at fostering its young. The bucks may get up to 275 pounds and does average around 125 pounds. They share many similarities to Boers with their hardiness but are considerably different in terms of how quickly they grow and how low-maintenance they are. Kiko goats are overall quite docile thanks to the history of being bred to dairy goats to produce their breed (Pieper 2021).

Myotonic

Myotonic goats are notoriously known for being "fainting goats." You'll see on many viral videos all across the internet people intentionally startling them so they will fall over, their legs stiff as a board. Unlike most goats, they are not known for jumping fences or climbing.

They have a peculiar origin story. The first known Myotonic goats were sold by a transient farm worker named John Tinsley to Dr. H. H. Mayberry's farm in Tennessee in the 1870s. Tinsley never told anyone where he had gotten the goats from and strangely vanished after the selling, leaving behind the first documentation of these goats ever to be discovered.

They are overall docile, curious, and friendly creatures. Though they have been given the adorable nickname of "fainting goats," the truth is they do not faint in the literal sense of the word. Instead, the muscle fibers of the animal lock up upon being overly excited or started, and they will fall over if unbalanced. This causes no problem to the goats, and they stay conscious for the entire duration of their "fainting" episode. If you cannot bear to think of sending these charming creatures to slaughter, take comfort that they can produce milk "on demand." They can be bred year-round or can simply be kept as a pet (AGF 2021).

If you do, however, wish to send them to the butcher, at the very least they will be incredibly easy to catch once they faint.

Savannah

Indigenous to South Africa and imported to the United States in the late 1990s, these goats are usually completely white. They bear a striking resemblance to the Boer breed in their musculature and personalities. Savannah (or Savanna, as the spelling may differ based on the market) are low-maintenance, fast growing, and known for their beautiful coats (AGF 2021).

TexMaster™

The newest breed of all goats on the market, TexMaster™ goats were only created and sustained 20 years ago to the date of this publication. You read that correctly—20 years ago, the Onion Creek Ranch began crossing Myotonic bucks with Boer does, and after several generations, they produced TexMaster™ goats. Though they are mostly Myotonic, the Boer bloodline helps them with their hardiness, musculature, and growth speed with the added benefit of being parasite-resistant thanks to the Myotonic strain (AGF 2021).

CHAPTER TWO:

Where and How to Buy a Goat

Before you rush off to the market to purchase your new goat, **STOP!** This may come as a shock, but not all goats are born equal—or rather, not all goats are going to be of the quality that you need for your chosen field of interest.

All too often, those that buy goats hoping for a lovely bit of meat or milk end up with a sickly animal that needs so much tender loving care that the goat never pays for itself with its output—or worse, it dies in the same week of purchase. If you're looking to rescue sickly goats, be my guest in finding the most diseased looking critter you can, but for those of you that hope to end up with the healthiest goat possible, take heed. There is a surprisingly political aspect to goat purchasing in regards to creed and lineage. Let's find you a prince and/or princess of a goat to rule your pastures in perfect health!

For more information about how to care for and identify sickness or injury in goats, please read chapter five.

TERMS TO KNOW
Before I launch into a spiel about what to do and know, here's some terms you'll hear frequently from here on out. Please refer to this section if you come across anything that confuses you. These terms are relevant to purchasing, maintenance, and easy-to-spot health issues of goats. The following terms are taken directly from the U.S. Department of Agriculture's glossary.

ADJUSTED WEIGHT: Weight of the animal that has been adjusted using a correction formula to a standard age, sex, type of birth/rearing, and/or age of dam. These weights increase the accuracy of comparisons between animals for selection as they account for known differences in environment. Adjusted weights are often used when doing performance testing.

ANESTROUS PERIOD: The time when the female does not exhibit estrus (heat); the non-breeding season.

ANTHELMINTIC: A drug that expels or kills internal parasites.

AVERAGE DAILY GAIN (ADG): The amount of weight gained each day during a period of time.

BACK CROSS: Breeding a first cross offspring back to one of the parental breeds. This is often the first step in establishing a grading up program or composite breed.

BLIND TEAT: A non-functional teat on the udder of the goat. It can be an additional teat that is not connected to a milk duct or one that is nonfunctional due to mastitis.

BODY CONDITION SCORE: A numeric value assigned to an animal that estimates the degree of fatness or condition that covers the animal's body. This score is assessed by palpating the spine (spinal and transverse processes) and ribs.

BOLUS: A rounded mass of medicine used in cattle, goats and sheep.

BULL DOG, UNDERSHOT or MONKEY MOUTH: The lower jaw is longer than the upper jaw, and the teeth extend forward past the dental pad on the upper jaw. This is a disqualifying feature for confirmation.

CALIFORNIA MASTITIS TEST (CMT): A kit that can be used to test mastitis in cows and does.

CAPRINE ARTHRITIS ENCEPHALITIS (CAE): An infectious disease that causes arthritis and progressive inflammation in one or more organs or tissue systems such as the joints, bursae, brain, spinal cord, lungs, or udder. This disease affects goats and is currently incurable.

CLOSED HERD or FLOCK: No new animals are introduced into the herd or flock.

CLOSTRIDIAL INFECTION: A bacterial infection that can occur in sheep and goats. Some goat diseases that are caused by this infection are Blackleg, Enterotoxaemia (Overeating disease), and Tetanus.

COCCIDIOSIS: A disease that is commonly exhibited in younger animals caused by a protozoa parasite infection. It is characterized by diarrhea, dehydration, weight loss, lack of thriftiness, and weakness.

CROSS BREED: An animal whose parents are of two different breeds.

CRYPTORCHID: A condition where one or both testicles fail to descend into the scrotum sac.

ENTEROTOXEMIA: A disease caused by an overgrowth of bacteria (Clostridia perfringens) in the intestine, usually due to fermentation of a large quantity of starch with production of endotoxin. Usually causes rapid death of animals.

EXTERNAL PARASITE: These parasites feed on body tissue such as blood, skin, and hair. The wounds and skin irritation produced by these parasites result in discomfort and irritation to the animal. Some examples of external parasites are: fleas, keds, lice, mites, nose-blot flies, and ticks.

FAMACHA: It is an acronym for Faffa Malan Chart; he is the person who developed a method of using the color of the inner eye lid to determine the severity of parasitic infection in sheep and goats in South Africa. The method is used to implement selective treatment programs for parasites in goats. To use the system properly, producers need to attend training courses and obtain an official chart. This system is only good for control of H. Contortus (also known as the barber pole worm).

GENOTYPE: The specific genes that the animal has on its chromosomes. The genotype of an animal is set at conception and controls the potential performance, color, size, and fertility of the animal. The genotype and environment combine to produce the phenotype of the animal.

HERMAPHRODITE: A sterile animal with reproductive organs of both sexes.

INBREEDING: The mating of closely related individuals.

INTERNAL PARASITES: Parasites located in the gastrointestinal system in animals.

JOHNE'S DISEASE (Mycobacterium paratuberculosis): A bacterial disease causing severe weight loss and some diarrhea. Not currently curable.

KID: A goat less than one year old.

LINE BREEDING: A form of inbreeding that attempts to concentrate the genetic makeup of some ancestor.

LUNGWORMS: Roundworms found in the respiratory tract and lung tissue.

MASTITIS: Inflammation of the udder usually caused by a bacterial infection.

METABOLIC DISEASE: Those diseases that involve the lack of or unusual breakdown of physical and chemical processes in the body. Often associated with nutrition and feeding.

OPEN SHOULDERS (Loose shoulders): The shoulder blades are structurally too far apart at the top, which makes it difficult for the animal to stand for long periods or to move around freely.

OVERSHOT or PARROT MOUTH: An animal that has a lower jaw shorter than the upper jaw, with the lower teeth that hit the back of the dental pad. This is a disqualifying feature for confirmation.

PARASITE: An organism that lives on or in another living organism (host) at the expense of the latter.

PEDIGREE: A listing of the ancestors of an animal that generally goes back 4-8 generations. It is often used to prove parentage for registration in a breed association. A shorter list can be used by producers to trace parentage of animals on their farm.

PERFORMANCE DATA: Information related to the growth rate of the goat. This often will include birth to weaning data and adjusted weaning weights. It correctly refers to any weight and animal gain data available on an animal.

PHENOTYPE: The visible or measurable result of genotype and environment. The phenotype includes an animal's external appearance, measures of its productivity, and its physiological characteristics.

POLIOENCEPHALOMALACIA, PEM, or "polio:" A neurological disease of goats caused by thiamine deficiency. The rumen normally produces adequate levels of thiamine; but under some conditions, such as a high grain diet, high sulfur in the diet, stress, or being "off feed," the thiamine is degraded, thus causing the disease.

PUREBRED: An individual whose parents are of the same breed and can be traced back to the establishment of that particular breed through the records of a registry association.

PURULENT: A term describing pus-like discharge or infection.

RECESSIVE GENE: A gene which must be present on both chromosomes in a pair to show outward signs of a certain characteristic.

REGISTERED: A goat whose birth and ancestry has been recorded by a registry association.

SICKLE-HOCKED: Condition when an animal has too much angle or set to the hock. This condition, when viewed from the side, is identified as the animal having their feet too far underneath them while the hock is in the correct position behind the animal.

SIRE: Male parent.

SKIN TENT: When the skin of an animal is gently pinched and pulled outward. A dehydrated animal's skin will not rapidly return to its normal position or shape.

SORE MOUTH: A highly contagious (including to humans) viral infection that causes scabs around the mouth, nostrils, and eyes and may affect the udders of lactating does.

TAPEWORMS: Long, ribbon-like segmented flatworms that can inhabit the gastrointestinal tract of animals.

URINARY CALCULI: A metabolic disease of males characterized by the formation of stones within the urinary tract. It is caused primarily by an imbalance of dietary calcium and phosphorus.

WHITE MUSCLE DISEASE: Problem in young goats caused by a deficiency of selenium and/or vitamin E. It causes kids to be weak at birth and shortly after birth. The condition impairs the animal's ability to transport oxygen properly and if not treated can result in death within 48 hours of birth.

YEARLING: A male or female sheep or goat that is between one and two years of age.

ZOONOSIS or ZOONOTIC: Any animal disease that can be spread to humans.

PREPARING TO BRING HOME YOUR GOAT

Before you go out shopping for your goat, you need to be sure you bring home your new animal and have a place to put it. After all, you can't keep a goat in your living room (unless that was the plan all along). Here are the basics of what you need to have ready for your goat before you make a purchase.

The first thing you should have ready is the habitat in which you expect your goat to spend most of their time. We go over this detail in depth in chapter three, but the bare minimum you need to have ready before you bring your goats home is somewhere for them to be where they will not escape easily—goats are notorious escape artists. Have a housing situation ready so that they can escape from the elements. Ensure you have a good amount of land the goats can explore freely within sturdy fencing—keep them in and keep predators out!

One of the most important things to have ready by far is feeding supply. Be sure to have feed storage containers, food bowls, a hay manger ready and full of hay bales, a mineral feeder, and water buckets. Goats need fresh water every single day, so see if you can't have a spigot right next to their water trough for ease of access. It's way easier than dragging a hose out where you need it to be! Be sure to have plenty of hay for the goats to munch on, grains, and minerals. If there are certain diets you want them to have, extensively research exactly how to get to the desired output you want from your goat.

In addition to this, have medicine ready for your goat! To see what you should have on-hand to care for your goat, please see chapter five.

TRANSPORTING YOUR GOAT HOME

Now that you've prepared to bring your goats home, it's time to consider how you're getting to-and-from the place you buy them from! Here's what to know about how to transport your goat.

Luckily, due to the scale of these creatures, there are lots of options to transport them that aren't available for most livestock. Possibly one of the easiest ways is to purchase a large dog crate to load the goat into before putting them in your vehicle. Closely related to this possibility is the aptly named "goat tote," an enclosed wire cage to sit in the back of a pickup truck specifically designed to house goats.

An enclosed pickup truck's bed would also be great for getting your critters home. Some owners have even used a minivan's trunk space to get their goat home! In that last scenario, farmers put bars up between the backseat and the trunk, then set down a tarp and covered it in wood shavings. The farmers in question from The Goat Mentor website recommend giving your goats probiotics before any trip home to help keep their gut flora balanced during the stressful journey.

If you have more than one to transport at a time, however, I would recommend investing in a horse trailer at the scale of your needs— consider the total number of goats you'll need to transport at any given time. Just be sure to have a vehicle with the correct amount of power for towing a trailer, or you can risk ruining your transmission (GoatMentor 2020).

QUESTIONS TO ASK ABOUT YOUR NEW GOAT

During purchase, always ask if the goat is registered and if they have had any vaccinations. Just like most pets, paperwork is necessary to ensure these creatures are well-documented or have some form of history at the veterinarian (unless they are very young).

Output is also very much tied to genetics! If you're hoping for a dairy goat, ask how much milk the mother produced regularly. If they're a meat goat, inquire about any issues the raiser had with their meat during the selling process in the past. Overall, be sure to educate yourself by speaking directly to the seller whenever possible.

WHERE TO BUY A GOAT

There are a lot of options for where to go when buying goats, some more involved than others. For buying in person, check local feed stores, which will often have bulletin boards up from local farmers. This is a great way to get acquainted with your local goat owner community, as well! Alternatively, if you hear of a local goat show, you can visit them to see who is selling, or go to a county fair that has a goat exhibit. Not all goats are for sale, though, so be sure you're in the right area before you offer money for someone's favorite pet!

If none of these options seem to be working out, then don't worry! The internet is here to save the day. To save on shipping costs, try Craigslist to see if there's anyone nearby for easy pickup, or even Facebook groups. Lastly, you can always go directly to the source by visiting a breeder's website, which will contain all the information you could ask for about the creatures you're interested in—albeit with the added cost of shipping (Smith 2021).

CHAPTER THREE:
Making a Habitat

There is an old saying in regards to making pens for goats: "If it can't hold water, it can't hold a goat." The majority of goats, regardless of breed, will be notorious escape artists. In this chapter, we will identify the different options of fencing or housing for these creatures, what you will need, and what type of materials to avoid when making your enclosure.

First and foremost, you will want a place that is indoors out of the wind and cold for your animals to hide in when the weather gets bad. This could be a barn with horse stalls that you let them into over the winter or on particularly hot summer days, or an insulated shack for them to go in and out of as they please.

In addition to this facility, you should have a pasture for them to graze and play in. Goats will not be content with sitting still; they need places to have fun and socialize with each other or humans. When in doubt, ask yourself: If I were a goat, what would I want to do in sunny, rainy, and snowy weather?

HOUSING

As mentioned before, it's vital that a goat has shelter to get away from the elements when it gets too hot, cold, or wet. Horse stalls in a barn are a fantastic option, but one can easily convert an existing shed to do the job.

Much like chicken coops, goat shelters are a common thing found on the market with a simple search. You can easily build one or buy one premade, just make sure you put it on high ground or somewhere not susceptible to flooding. Light and air should be easily accessible inside the shelter, and keep it dry inside as much as possible—damp conditions easily lead to disease. If you intend to allow your goats to breed, there must be a separate area for kidding where the younglings will not be exposed to the rest of the herd and the mother can be undisturbed. There needs to be at least a few inches of bedding atop a solid foundation for the goats to lay down on with a separate area for feeding and water. This is because any fecal matter dropped into water or feed could lead to the growth of parasites in the body, which is definitely something best avoided! The scale of this structure needs to be at least 10-15 square feet per mature goat, so assuming you have five mature goats, you would need a minimum of 50 square feet inside for them all to rest comfortably. See the list below for options on what bedding to use for your goats (Amherst 2021).

- Pine Shavings
- Straw
- Pellets (compressed wood)
- Sawdust
- Wood chips
- Cedar shavings
- Sand

Invest in a wheelbarrow and pitchfork—plastic is better, in my opinion—to regularly scoop manure out from the bedding. The last thing you want is a buildup of feces in the bedding as this can lead to the transmission of parasites and diseases.

As for what to do with the scooped poop, there's some good news. Goat poop is a fantastic fertilizer as, when a goat is healthy, it comes out as nearly odorless pellets that don't attract flies. Farmers will use this manure for their vegetable gardens, flower beds, and even crops due to its richness in nutrients and how good it is for the soil. My favorite thing to do with it is to use it for composting, as it takes away the "ick" factor some might have about using it as fertilizer directly, instead turning it into highly fertile soil over time.

Dump any ruined bedding from urine or manure into a compost pile you create nearby the shelter, but make sure it is outside the fencing so your herd doesn't go digging through it (Tilley 2019).

FENCING

A bare minimum of at least 25 square feet of land per animal is recommended for goats to trot around outside. A general rule is 12 goats per acre, so plan to have this much area available when creating the layout of your goat enclosure.

When considering what kind of fencing to use for a goat, remember: if it can't hold water, it can't hold a goat! In this case, however, you'll find that lots of the fencing recommended utilizes wire commonly found in poultry enclosures: chicken wire. While plotting what kind of fence to build or use, also remember that some goats will try to dig their way out or leap over fences to come say hello to you.

All goats will jump over fences shorter than four feet high, but depending on the breed size, you may have to scale it up to at least five feet tall. Goats may love to nuzzle their noses out of fences for a pet, but as cute as it is, it can be very dangerous! Should a horned goat stick their head out, they risk getting stuck, which can panic the animal and lead to serious injuries while trying to escape. Any gaps at all between posts, cross-braces, or the squares of a wire panel should be no larger than four by four inches.

Naturally, goats can only push, not pull. For added security, ensure you have a gate that only opens inward into their pen and not outward to freedom. The same goes for where the wires should be attached; put them inside toward the goat and not outside, where they can be pushed away from the fence posts keeping their enclosure up.

No matter how much planning you do, however, it is vital that you

maintain security by regularly walking along the fenceline to inspect it. Any problems like sagging or weak points formed by the goats can be caught early and mitigated before your entire herd is running loose (Everett 2021).

Options for fencing material include but aren't limited to:
- Wood
- Electric
- Woven goat wire
- Cattle/Goat panels
- Chain link fencing

All these options are very different in assembly and price range, so research between these options and see which one may be right for you.

PLAYTIME FOR YOUR GOATS
One last important thing to know about the enclosure that isn't specific to breed or whether they will be dairy or meat goats is that all of them love to play. They are generally very intelligent creatures that need mental stimulation from human interaction, exploring, and the topic of this section: toys!

Goats are not at all picky about toys you give them and will be happy with anything that provides jumping, noisemaking, or general stimulation. An online search for goat videos will often yield imagery of goats jumping onto play sets in the middle of a field for good reason: their need to play must be fulfilled for goats to stay happy! Just be sure any toy or obstacle placed in the pasture and/or in the shelter is at least five feet away from the fence, unless you want your goats flying out of the pen and onto the other side!

There's no limit to what goats will love to play with: old tires, seesaws, slides, dollar-tree items, stumps, old hanging cow bells—some farmers even build full obstacle courses for their goats to play on! Just make sure the space the goats can hop up onto is wide enough for them to stand on. They're very agile and elegant with their balance (aside from the myotonic, who "faint" when excited), so you may be surprised at what they can jump onto easily.

CHAPTER FOUR:

Feeding Goats

You may have heard that goats will eat just about anything, and that's not wrong, per se—it's just that they shouldn't eat just about anything. While some of them will be perfectly content grazing on what grass there is in your fields (and they do make excellent lawnmowers), others will have specific diets you should consider when thinking of their output.

I remember once, my mother brought home from a friend a supply of freshly butchered beef for our freezer. It was a very generous offer, but once we cooked and sampled the steaks, the meat tasted utterly rancid! Nothing was amiss and the meat was great quality, which left us scratching our heads as to what could have happened. As it turns out, the cows were in an apple orchard and would eat the fallen apples daily. As a result, their meat had a horrible taste. This is odd when you consider that apples would make the meat taste sweeter in the cooking process, but it instead gave the opposite effect when eaten by the cattle. This fascinating revelation helped to put into perspective what I should have my goats eat or not eat. Whatever I feed them will change the flavor of their meat or milk—not to mention how it affects their health, which is also very important to consider.

If you need more proof of what an animal eats affects the flavor of its output, look no further than your local grocery store. Most cattle and chickens are fed the same corn-based diet to fatten them up as fast as possible and keep their meat the same, consistent flavor. You may have heard of Kobe beef, which is meat from Kobe, Japan, a place in which famers carefully breed, care for, and feed cattle that will eventually be turned into the most famous cuts of meat—for good reason! The meat, without any seasoning, is one of the most flavorful in the world.

So, when you go to read this next chapter, ask yourself: do I want the milk and meat I make to taste good? The answer is yes, yes you do.

BASIC GOAT ANATOMY

Without diving into a large amount of medical terminology, some of the most essential knowledge you'll need to know is that goats have four stomachs. The first stomach will break down hay or feed eaten, which is then sent to the second stomach to be partially regurgitated into a cud. The goat will then chew this regurgitation and swallow it back down again. This is where the term "chewing your cud" comes from. Any accidentally eaten bits of trash or rubble will also be regurgitated with the cud and spat out, which is nice to know as a defensive mechanism for farmers worried they can't find their keys. (Just kidding!) Rumination is the name of this process, and it causes a goat to produce belches. A goat that is not emitting belches frequently can be subject to bloat. The final two stomachs will process food as it would normally to be passed as fecal matter after all nutritional value is absorbed.

While not relevant to the digestive system, it is also important to consider hoof care. A goat's hooves must be regularly trimmed so as to not overgrow and produce discomfort in the goat. For them, their hooves are like our fingernails: with regular care, they're nothing to worry about, but left on their own, you could end up with an uncomfortable pain in your feet every time you take a step.

Trim your goat's hooves a minimum of once per month and pick out any rocks or mud stuck in them. Imagine being stuck with a rock in your shoe, and you don't have any hands to pick it out! It's just another step in keeping your herd happy (Smith 2021).

NUTRITIONAL NEEDS FOR DAIRY AND MEAT GOATS

Dairy

When creating a "menu" for a dairy goat, the key to making sure they eat well all year is feeding them lots and lots of hay. Basic straw bale hay will work just fine for feeding most of the time, but during the high milk-production seasons, swap in legume-rich alfalfa bales. While milking or pregnant, feed your producing goat at least one portion of grain a day to keep them nutrient-rich with all the added vitamins and minerals. The grain combined with the alfalfa on top of the regular portions of hay will provide plenty of protein for your happy, jumping critters.

Likewise, when baby goats, kids, stop suckling from their mother after around two weeks old, they will begin to graze on the pasture like the rest of your herd. Eventually, they will take an interest in the hay and come to eat with the others. While they're growing, however, be sure to offer them a portion of grains to help them boost their immune system with vitamins and minerals.

Meat

Meat goats overall don't require much nuance to their diet; sticking with hay year-round, letting them forage, and keeping their water topped up is enough to get them through the year. Keep what they eat to be plant-based as much as possible, aside from the occasional grain bucket. Protein will be essential to their regular diet, so add supplements to their grain on occasion.

It is the goal of most farmers to fatten their herd up as quickly as possible to get the best market value. To raise the overall weight of your goat without damaging their health, there are two methods you can use. The first method is known as the traditional way: providing your goat with lots of fodder to eat at any given time, essentially ensuring they overeat and thus put on plenty of weight over time. The second method is by using agro-industrial by-products such as cotton seed, ground nut, and palm kernel cakes, or protein sources such as molasses and rice bran. This feed should be used in addition to the usual hay and forage-based diets the goats have. These products are usually quite inexpensive, so inquire about them at your local feed store for purchase.

WATERING YOUR HERD

Goats must have a constant source of clean water that is changed daily. Place the watering trough somewhere near the shelter where the goats can have easy access to it whenever they like. Rather than hauling buckets back and forth to refill their water receptacle, consider placing their troughs or buckets nearby a spigot or, at the very least, within reach of a hose. If necessary, I would recommend putting in a new waterline near the shelter.

Goats need a bare minimum of one gallon of water per day per goat, with more being added when considering the weather. If it's hot, four gallons or higher is expected per goat. If you expect to be gone for a few days with no one to fill your troughs or check them, look into getting a stock tank, which will automatically refill the water troughs with water you've already topped it up with. This will still require refilling, but this method can help with planning ahead if you're going to be gone for a short while.

If you're lucky, you may even have a creek or a pond as a frequent water supply for your livestock to drink from. Unfortunately, though, unless you have running water going through your pasture at all times, there is a risk that it can become contaminated with urine or manure, or even become stagnant and rife with bacteria. Have this water tested before allowing your goats to drink from it.

Likewise, if your water source is from the city, try talking to a water supplier about switching over to well water. City water is often full of chemicals to help purify it from any additives of pipes, whereas well water will generally be more pure and untouched by chemicals that would affect your goat's health (Smith 2021).

CHAPTER FIVE

Caring for a Sick or Injured Goat

While raising your livestock, you will over time come to understand their personalities and general behavior. It's when your critter is visibly not acting like themselves that you need to start being concerned.

Give your goats a general once-over at least weekly, if not every day. Make sure they have no visible injuries, and if there are none but they still don't seem well, separate them from the rest of the herd and keep an eye on them.

Putting them in the shelter or in an area they cannot leave is ideal in most situations when they're sick or injured to avoid the risk of exacerbating the issue. As mentioned in chapter three, it's good to have a separate area for allowing your goats to give birth. The same is true to quarantine or isolate a goat that is unwell until they seem better or are no longer at risk of further hurting themselves. Be cautious with this since isolated goats still need socialization and will do whatever they can to get it—provided they're healthy enough to want it at all.

FIRST AID FOR GOATS

Much like humans or livestock, goats can be susceptible to injuries or illness. For this reason, have a first aid kit ready to go in an easy-to-find place in case your goats seem under the weather the next time you check in on them.

If your goat ever has an accident or was butted too hard in a play-fight with its friends, the last thing you want is to be caught with your pants down with no medicine or bandages to provide. Although you should see a vet for any major injuries or illnesses that don't go away after a few days, here are a few things you should have on hand in case of emergencies.

First Aid Equipment for Goats	First Aid Medicine for Goats
• Surgical gloves	• 7% Iodine
• Drenching syringe for medicine administration	• Terramycin eye ointment
• Cotton balls	• Disinfectant Spray such as Hydrogen Peroxide
• Gauze bandage	• Blood stop powder
• Alcohol prep wipes	• Epinephrine
• Elastic bandage	• Antibiotic ointment
• Digital thermometer	• Activated Charcoal product in case of poisoning
• Tube-feeding kit for young goats	• Children's benadryl syrup for congestion or breathing issues
• Small clippers	• Aspirin
• Cotton-padded PVC pipe cut in half	• Copper sulfate or Zinc sulfate
• Vet Wrap	• Biomycin for pneumonia, pinkeye or infections
• Ice pack	• Milk of Magnesia for constipation or bloat

You can find almost all of these materials at your local feed or drug store, if not through online ordering.

Though not necessary for emergencies, consider also having Betadine surgical scrub for cleansing wounds, probiotics, powdered electrolytes for dehydration, hydrogen peroxide for cleansing wounds and rubbing alcohol to sterilize equipment before and after use (Smith 2021).

This may sound like a lot, but livestock are just as susceptible to disease and injury as we humans are, and they have about the equivalent need for care for when under duress.

Here's a tip for you: if you're ever in a pinch, and you need to cover a wound on your livestock until you can get to your first aid kit, have women's maxi pads on hand! You'd be amazed how great they work as a makeshift bandage, since the bottom side is sticky and the interior is layered with cotton for absorbing blood. Plus, they are generally individually packaged and ready to go at a moment's notice in case of emergency. Depending on the size of your goat, a medium size should work just fine, and they're less expensive than traditional bandages for livestock. In emergency situations, I've seen my mother cover our horses' wounded leg with one she had on hand, and ever since we've always taken them with us while working with livestock or trail riding. Food for thought!

HOW TO USE YOUR FIRST AID KIT FOR COMMON INJURIES (AND WHEN TO GO TO THE VET!)

Wounds, Cuts, and Scrapes

At first it may seem very alarming to find your goat friend squalling and screaming, but don't be fooled. As fun as goats are, they can be noisy little drama queens over tiny injuries. These are usually minor and not much cause for concern, but it's best to treat them as soon as you find them to prevent complications.

First, get your goat on a halter or on a stand—just make sure he's standing still. Carefully use your clippers to cut away the hair surrounding the wound, then clean the injury with a little water to clear away dirt. Next, spray the area with your disinfectant spray to clear away the bacteria or rub with the antibiotic ointment. Pat the area dry and wrap with bandages (Garman 2021).

If the top of the horn is broken and not bleeding, it cannot be mended. It is possible to smooth out with sandpaper if you wish, but it is usually not necessary.

The moment you see blood, immediately find your blood stop powder, dress the horn, and go to the vet. I would recommend taking advice directly from them.

As you become a more seasoned goat owner, you will be able to treat some of the less intensive breaks on your own. However, if the base breaks to the point where it's down to the head, it does not matter how confident you feel.

A horn broken at the base of the skull is an emergency. Immediately cover the open hole on the head, which again leads to the sinus cavity, with gauze, vet wrap, and bandages and get to the vet quickly as you can. If in the event a vet is not available, clean out the cavity with a sterile solution to remove debris, such as saline water. Then cover with gauze and vet wrap. Change the dressings every few days and do not leave the area unprotected until the hole closes.

During the healing process, keep the goat in an isolated area to minimize stress or risk of further injury.

COMMON DISEASES AND HOW TO TREAT THEM

Bloat

Bloating happens when an excessive amount of gas is trapped in the rumen. Bloat can be divided into two types.

Frothy bloat is common and occurs when the goat consumes an excessive amount of moist hay or lush pasture, especially in the spring. The second type of bloat is free gas bloat. If there is a solid obstacle to eructation (such as an apple), gas is held in the rumen. It can happen as a result of a protracted duration of lateral recumbency. That means that fluid inside the rumen blocks the outlet where gas exits (rumen cardia). Both conditions are extremely dangerous and must be treated immediately. A distended, rough stomach is a symptom, as are the animal's stomping feet, loud bleating, and stiff-legged walking.

Coccidiosis

Coccidiosis is caused by a parasite that is found within the cells of a goat's intestines. Depending upon the number of parasites present in the intestines, it can dictate how severe a case may be. If this animal becomes stressed, you will notice the symptoms become worse and lower the animal's resistance to other diseases. Symptoms include bloody diarrhea, a loss of appetite, weight loss, and a possibility of sudden death. Treatment is the sulfa drug, isolating the sick animal, and good sanitation practices on the property.

Flystrike

Flystrike is fairly rare in goats, as the primary target is typically sheep. This occurs when maggots of blowflies hatch on the skin and feed on the animal's living tissue. This occurs when adult flies lay their eggs in a moistened coat from urine or fecal staining. Skin wounds, weeping eyes, or lesions from footrot can be targeted by the adult flies. Symptoms will include agitation, secondary bacterial infections, picking at infected areas, exhibiting a foul smell, and flies targeting one particular animal. Treatment for this includes close clipping of affected areas, removing the affected animals from the balance of the herd, cleaning and dressing any area showing infection, and the application of antibiotics at the discretion of your veterinarian.

Floppy Kid Syndrome

This disease is most common in kids 7 to 10 days old and is associated with overfeeding. Do not permit the kid to drink any more milk and use baking soda and Milk of Magnesia to get all of the ingested milk out of the system while lowering the acidity in the digestive tract. Administering C&D antitoxin will help.

Mastitis

A disease of the udder caused by staphylococci, streptococci, E. coli, or mycoplasma agalactiae. Watery/chunky milk, blood-tinged milk, a solid, hot udder, and reduced milk production are all symptoms. Mastitis may be either acute or chronic.

Goat Ketosis

A metabolic disorder caused by an incorrect feed balance. If untreated, it often progresses to enterotoxemia (see below), which is fatal. Depression, general malaise, going off feed, staggering, heavy breathing, and teeth grinding are all symptoms. The most obvious sign that a goat is in ketosis is that her urine smells sweet and fruity due to the presence of excess ketones.

Enterotoxemia

Also referred to as the overeating condition, enterotoxemia is triggered by toxic amounts of Clostridium perfringens type D (CD), usually present in the rumen's natural flora. Under some conditions (including excess grain, excessive milk, insufficient roughage, or high parasite load), the CD reproduces rapidly within the rumen, causing acidosis. Without prompt treatment, death can occur unexpectedly. Watery and/or bloody diarrhea, abdominal pain, lack of appetite, and subnormal temperature are common symptoms at the onset.

Pneumonia

An inflammation or infection of the lungs caused by viruses, bacteria, parasites, mycoplasmas, fungus, irritating fumes, dust, or moist environments. Pneumonia can be prevented by keeping your goats and their environment clean.

Tapeworm

Tapeworms are one of two parasites that can be seen in the feces, making them easily identifiable. Adult worms are transparent, ribbon-like, and/or segmented in the manner of rice grains. Eggs of tapeworms are small and triangular. Although tapeworms are generally harmless to adults, they can block the digestive tracts of young kids, resulting in complications and death. For treatment, oral administration of fenbendazole (Safeguard, Panacur) at a quarter of the dose recommended for horses/cattle is recommended.

HOW TO AVOID DISEASES

Don't forget to keep a close eye on your goats every day, looking out for symptoms of illnesses—and remember that, no matter what you do, it is still possible for your goat to die suddenly.
If this happens, do not torture yourself with guilt or what-ifs. No matter how vigilant you are, you are neither omniscient nor omnipresent. You can't catch everything. In case this ever happens, your best bet for finding out what caused sudden death is to take a fecal sample to the vet to find out the cause of the death. This way, you know what particular symptoms to watch out for next time.

Clean your barn or goat shelter daily. Goats defecate very often, so you need to clean this up daily to keep their home clean and sanitary. Remember that goats drink plenty of water, so you will need to refill their water buckets daily. Make sure that you are providing about 12 liters of water per goat. This is a little on the high side but will make up for any spillage and ensure that your goats have more than enough to drink. A dehydrated goat is a stressed and unhealthy goat. If you have dairy goats and are not providing them with enough water to drink regularly, they will, in turn, produce less milk than average.

Make sure that your goats drink clean water so as to stay in excellent health. To ensure this, pour out and refill their buckets twice a day. If you cannot, perhaps because of work or family commitments, then once a day is also acceptable, as long as you change their water on a day-to-day basis. Of course, if you have a crick in your goat's pen, it is easier to keep the water supply replenished so long as it isn't polluted.

DAILY CARE

Cleanliness is a must in all areas of goat care. Work hard to keep your livestock and their surroundings clean at all times because it reduces cases of infections and diseases, saving money in veterinary care and time spent restoring a sick goat or herd back to health.

After cleaning up waste from your goats every morning, clean up the soiled and damp straw. Regularly provide your goats with plenty of clean, fresh, dry straw to keep them comfortable and happy. You may also use hydrated lime on wet areas in the barn. Hydrated lime works to keep dampness and moisture under control and to prevent the spread of bacteria. Hydrated lime also helps to clean up the ammonia from goats' urine. This is important to note because ammonia in confined spaces, even in low concentrations, can be dangerous to a goat's respiratory system. It is also hazardous for humans to breathe.

However, in recent years, many farmers have switched to natural ammonia control products because hydrated lime has been proven to produce more ammonia gas when used on livestock urine (Absorbent Products, 2019). There are plenty of natural ammonia control products on the market, and if you can not find them locally, they are easy to find online. Wendy Powers (2004) from Iowa State University writes that there are alternatives to reducing the amount of ammonia in your barn or shelter. One such method is to use biofilters that trap emissions and also "provide an environment for aerobic biological degradation of trapped [ammonia] compounds."

The favorite all-natural barn refresher is Sweet PDZ. Choose to use all-natural products because goats have susceptible lungs, and they will quickly become ill if they are kept in a dirty barn. Furthermore, you do not want the ammonia in their urine to turn toxic, and it has been found that Sweet PDZ mitigates all these problems for everyone involved. It is also straightforward to use because all you do is sprinkle a layer of the product on the floor before topping it with straw and bedding.

Keep an eye on your goats; they are like children, and you must watch them consistently for any changes in behavior or mood or any unexpected physical changes. The best time to do this, of course, is during your daily care routine.

Watch out for any goat that is not eating much or that is not moving much. Other symptoms that signal a health problem are limping or crying, difficulty breathing, a high temperature, coughing, shivering or a mother goat not feeding her kids. Vomiting, diarrhea, or producing discharge from their nose or eyes are signs of a serious problem.

Also look out for any scrapes, cuts and wounds. If you notice any, you immediately rub the area with some iodine to disinfect the site and keep it clean of infections. Since goats love playing all day, expect to see scrapes, cuts and wounds regularly. You should go so far as to inspect your dairy goats for abrasions while milking them and then inspect the other goats during feeding time.

If you notice any of these symptoms, the first thing to do is gently check the sick goat's rectal temperature using a digital thermometer. (Goats do not like having their rectal temperature being taken.) You can purchase digital rectal thermometers at many supermarket stores or farm stores. An average temperature for a goat is between 101 and 103 degrees Fahrenheit. If your goat's temperature runs higher, it is time to call a veterinarian immediately.

Before taking the goat to the veterinarian, collect a handful of fresh fecal samples in a clean container or bag. This makes it easier for your vet to figure out what is wrong with your goat. Your vet will be able to test the fecal samples and tell if your goat has the following diseases/health problems:

- Overeating
- Stomach problems, including E-coli
- Parasites, including coccidiosis

Essential Supplies

If you cannot reach your vet as soon as you would like, you can pack some essential supplies that will help keep your goat in good health or stable condition until you can see your vet. Here I've put together a helpful list of things you may need in medical emergencies if you're caring for a sick goat. It includes:

- A Drench Gun (for administering dewormers and meds)
- Vitamin B-12 Complex
- Sulfa Meds (for coccidiosis)
- Dewormers (use the same dewormer for that worm until it no

longer works)
- Antibiotics: LA200, Nuflor, Penicillin, or other antibiotics recommended by your vet.
- Skin moisturizer.
- Probiotics
- Wound spray and wound dressing.
- Syringes with needles
- Disinfectant (Iodine)
- Anti-Inflammatories
- Red Cell or Iron for Anemia
- Immune Boosters
- CD-T Toxoid: Long-Acting Booster
- CD-T Anti-Toxin: Immediate Action

Vaccinations for Your Goat

A vaccine is a product that stimulates the immune system. There are several types of vaccines, including live or attenuated, dead or inactivated, and toxoid. There are others that are at an experimental stage and will not be covered in this book.

A live or attenuated vaccine actually has live microorganisms that cause a certain disease. These microorganisms have been modified, of course, so that they cannot cause the disease in the patient. They're still potentially dangerous to use, especially in the case of diseases that can also infect humans. They also tend to invoke the best, most protective immunity.

A dead or inactivated vaccine, as the name implies, contains dead microorganisms. It is safer to use, albeit provides less immunity, thus the need for various vaccinations. A live vaccine requires only one shot, while an inactivated one usually requires two or three shots with a month interval between.

A toxoid vaccine is one that is produced from a compound that the microorganism produces rather than a micro-organism itself. The most famous is the vaccine for Tetanus.

In the past, it was often recommended that goats be vaccinated every year against all diseases. This thinking has changed in recent years. Now, it is recommended that vaccinations only be applied as necessary.

It is known today that the body has a limited capacity to respond to vaccines. This means that only a few vaccines can be applied together at one time. Applying too many vaccines at one time will be throwing money away, because the body is incapable of producing antibodies against all of the diseases at once.

It is best that vaccine titers be taken before vaccination. This is a blood test that measures the number of antibodies for that disease in the goat's blood. This is normally not done because of the associated costs. Instead, the pertinent vaccines are given annually. Your goats will need to be vaccinated for rabies, tetanus and clostridium (CTD) with booster shots for tetanus annually. Give your goats the other vaccines one to two months before the rabies vaccine because the rabies vaccine affects goats badly. You do not want to overwhelm your goats' system by giving them all three vaccines simultaneously.

Goats are friendly, intelligent, and fascinating animals and it's critical to understand what to expect if you plan to keep them.

CHAPTER SIX:
Butting Heads and Joyful Jumping:
UNDERSTANDING GOAT BEHAVIOR

THE FLEHMEN'S RESPONSE

When the goat twists his or her upper lip, extends his or her neck, and appears to smell the air with his or her tongue, this is the Flehmen Response. This response may be shown by male and female goats of any generation. Many other species, including horses, donkeys, giraffes, cats, cows, yak, sheep, and rhinos, exhibit the Flehmen Response. When the goat detects something and must investigate more, the response is activated. Like all other animals that exhibit the Flehmen, Goats have an olfactory sensory organ known as the Jacobson's organ on the mouth roof that allows them to "smell" and sense pheromones.

The Flehmen Response is simply a type of self-communication. When goats detect something they are interested in, they will use the Flehmen to collect details about the smell. Urine and birthing fluids are often examined with the Flehmen and the hind end of a doe to determine the estrus stage. The outcome of their investigation offers an answer to the goat's query, "What's that smell?" We will discuss the Flehmen Response in greater detail in the following section.

PAWING AT THE GROUND

To Attract Your Attention: Goats, especially spoiled infants and adults who were spoiled infants, will paw at your legs to attract your attention. This action is often followed by nibbling on your clothing or fingers. This means "TREAT, PLEASE!" or "PET ME IMMEDIATELY!"

Hormones: When a buck cannot approach a doe through the fence, he will paw the ground in frustration. This is often accompanied by excessive lip and tongue blabbering and vocalizations. As part of their pre-mating practice, you may see a buck kick out its legs. Does in labor paw the ground in preparation for kidding, while does that are in heat will often paw the ground and act "bucky."

Climbing and Having Fun
Goats are inherently curious creatures, and goats of all ages love climbing. I have many spools, stumps, bottles, shelves, and pallets of varying heights in each of my goat yards. Also, we have a plethora of Little Tykes slides, houses, cubes, and tables for the goats to investigate. It's critical to have different games to avoid boredom and because a goat will often choose to nap on an elevated shelf or platform over the ground.

Babies are incredibly energetic and constantly bounce, skip, and run around. If the herd is in an especially cheerful mood, you can see both old and young bouncing down the hill to the pasture, running their feet out from side to side and jumping. Knowing your goats are happy and safe is about the greatest feeling in the world.

HEADBUTTS
Goats headbutt to demonstrate their superiority over other goats. The process of acquiring a place in a herd fascinates me, and thus, this section will be lengthy—but hopefully worth reading.

To start, every herd, whether it contains two or a hundred goats, has a hierarchy or official status for each goat. Although the goats are the only ones who fully understand their position, humans will observe and note it if they watch long enough.

When a goat leaves or is introduced to the hierarchy, there will be problems in the hierarchy. Do not be concerned that they will cause harm to one another. It's **normal**! And, in most cases, they do not draw blood. Occasionally, someone can strike another doe hard enough to knock a scur off, resulting in a little raw and/or bloody action. There have never been any serious headbutting injuries in my herd, but all of my goats are hornless.

When your herd becomes large enough, you may notice small family groups and cliques forming within it. It's incredible to see babies lying alongside their mothers, grandmothers, and great-grandmothers. Having two or three goats with no family in the herd band together and form their clique is equally amazing. Goats are truly remarkable and fascinating creatures, and they are *so much fun* to observe!

ADDITIONAL INTRIGUING GOAT BEHAVIORS

Goats are highly intelligent creatures who are driven by food. It is possible to teach goats different tricks and commands.

Queen Mary University of London researchers once taught goats to solve a puzzle intended for primates with impressive results. They replicated the puzzle ten months later with the goats, and the goats performed much better. Naturally, goats also possess remarkable memories.

You can train goats to respond to simple commands such as "get up," "get down," and "come." When I milk, I always milk in the same order each day. I have eight things in my milking this year. Chai is the first to arrive at the gate to be let into the milking parlor. Dido approaches the gate to be let into the milking parlor when she hears the stanchion unlatch.

I reaffirm the order by addressing them by name, assuring them that their turn has come to be milked. Their incentive is food, as they are permitted to consume as much grain as they want while being milked. Precious is followed by Diva, Jessie, Arianna, Cleopatra, and Delilah. They are all familiar with their assigned order, and they all come to line up each morning and evening. Diva likes to eat from the wrong side of the stand, so she normally circles around the front and munches before I say, "Diva! Get up!" and she leaps up onto the stanchion to be milked. Goats rarely expel gas from their rear ends, but they often burp, especially when chewing their cud. When goats sense threat, they sneeze to warn the rest of the herd.

While I have no personal experience with potty training goats, I know many breeders who have done so quite successfully. I don't believe it would be difficult to train a goat to pee in a specific location, but I doubt training them to poop there is very effective. Poop tends to come out at the most inopportune moments, wherever the goat is.

Recently, I've been supplying goats to a nearby yoga studio for Goat Yoga, which allows the goats to communicate with humans and demonstrate more of their ridiculous behaviors. It was previously mentioned that goats enjoy climbing and being higher than the ground. Yoga is an ideal environment for goats to hop on yogis' backs and frolic around the studio. It's a wonderful time filled with laughter. If a goat hops on your back, you should be honored, as goats communicate in this manner. You are a member of my family.

PHYSICAL & LIFESTYLE CHARACTERISTICS OF GOATS
Understanding goat characteristics is critical for effective goat farming. Goats are typical domestic or farm animals. Also, they make excellent pets.

In nature, there are two types of goats. Capra hircus, the domestic goat and the wild goat (Capra aegagrus). However, all domestic goats are descended from wild goats. Wild goats are found in eastern Europe and parts of southwestern and central Asia. Although wild goats are similar to domesticated goats, they have distinct characteristics.
Here we will discuss the physical and behavioral features of goats.

Physical Characteristics of Goats
Many domestic goat breeds are available worldwide. On average, domestic goats weigh between 15 and 130 kilograms, depending on the breed. The coat color varies according to breed.

The most seen colors in domestic goats are black, brown, red, tan, and white. Also, some goats have different colors on their bodies. The majority of domestic goats are horned and have beards. Male goats are usually larger than female goats.

Characteristics of the Goat's Lifestyle

Many domestic goat breeds are available worldwide and many farms raise these goats. Domestic goat management is very easy. You can easily handle goats if your farm has an adequate amount of grass or other greenery.

Domestic goats are very friendly and gentle by nature. They are relatively easy to raise due to their docile disposition. If you confine your domestic goats inside a predator-free fence, they can thrive and grow there.

By nature, goats are grazing animals that consume many greens and grass (both fresh and dry). Domestic goats also enjoy eating different leaves and grains. Also, they enjoy living in a clean climate.

CHAPTER SEVEN:
Caring For and Grooming Your Goat

Goats are not too difficult to take care of, provided you keep their housing clean and free from poop and pee. They need access to good nutrition, clean fresh water, pasture area, sunlight, and a cool breeze— this should do a lot to keep your goats in good health. Still, you'll need to groom your goat from time to time.

Goats are skittish creatures and can be nervous, so it's best to be calm around them; be quiet and gentle with them. You can buy goat rope halters to lead goats to where you want them to be, and if you need to handle the goat for any other reason. Having another person to help can be useful, so the goat doesn't become distressed and start to struggle. The more you handle goats whilst they're kids, the more they'll be used to being handled as adults, making your job much easier.

Routine health care costs will include pesticides to prevent mites or lice on your goats (approx. $20). You can get Pyrethrin powder to rub on goats on their back, or Ivermectin Pour-On for cattle. You can buy ophthalmic ointment to treat pink eye ($20) and dewormers (again about $20).

In order to adequately groom a goat, you need a curry comb, hard and soft brushes, a comb for beards and tails, and hoof trimmers. If your goat is mucky, then a bath mitt can be useful too. Electric clippers can neaten up their tails or general appearance.

BRUSHING

You can get rid of any mud on the goat by brushing it with a hard brush; if there are less obvious gritty bits in the goat's coat, then the curry comb will bring these out and give the goat a massage, too. The soft brush should be used to finish off the grooming, and it will distribute natural oils through your goats' coats.

Grooming your goat also allows you the perfect opportunity to check on the health of your goat. Brushing your goats is a good time to also feel their bodies with your hands and look for any bumps and lumps. These could be a parasitic skin infestation or wounds. Be careful and don't brush your goats too hard in wounded places.

You shouldn't need to bathe a goat, but if your goat has parasites, this may be an exception to get rid of lice and help with clipping a goat. The water should be slightly warm when bathing a goat. Washing a goat is not difficult at all; just wet the goat, use an animal or goat shampoo, and rinse.

HOOF CARE

Goats' hooves should be trimmed every 4–6 weeks, and I can't stress enough to set up a **firm schedule** for when this will be done to maintain them. You can keep your schedule on a regular calendar, in a diary, or on an app on your phone or computer. If hooves aren't looked after, goats can become lame or their hooves can contract an infection.

If you start the practice when they are young, they will be less resistant to it, but they are still not keen on having their back hooves done. You can get a person to stand at their front to distract them with a treat while you tend to their back hooves.You can also use a milking stand to keep the goat in place whilst it's having its hooves trimmed.

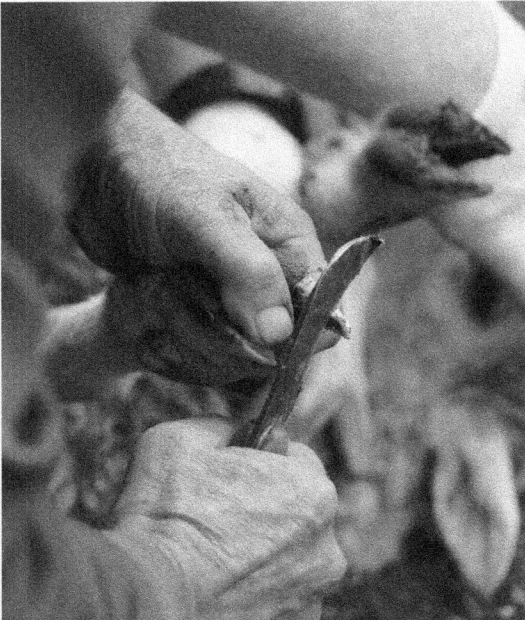

If you feel confident enough, and have a number of goats, it would be beneficial to learn how to trim your own goats' hooves with some $30 trimmers, rather than repeated expensive trips to the vets for this. A vet should be able to show you how to do this. If possible, it's always best to learn from a veterinarian or an experienced goat-keeper. There are nuances in trimming technique that cannot be conveyed through words alone.

Before you get to trimming your goats' hooves, it can be useful to have some corn starch handy in case you accidentally trim the hoof too close and cause it to bleed—you can apply corn starch and a dab of antibiotic ointment. It's sensible to wear thick gloves when hoof trimming, as the clippers are sharp, and the animals will resist and make movements. When you are taking care of hooves, ensure there are no rocks or sticks stuck in the goat's hoof, and also check that their hooves don't smell, as this can be a sign of hoof rot. As for the trimming procedure, it's not too different in concept from trimming a dog's nails. Just like cats or dogs, goats have a sensitive area made of soft tissue in the center of their hooves, which is called the quick.

Once your goat allows you to raise their foot relatively calmly, use a brush to clean any surface dirt off their hooves. Then, take a clean pair of hoof trimmers and use the tip to clean out the dirt from your goat's hoof, especially between the hoof wall and the soft sole of their feet. It's highly likely there will be a lot of dirt and debris to dig out.

I recommend that you start trimming at the very front of their toes, but it depends on how overgrown the hoof is. Start clipping the overgrown part of the toe tip just a bit at a time. As you keep trimming the hoof, you'll see that the surface of the remaining hoof wall starts to turn white (or black, if your goats have black hooves). Always trim only a little bit at a time. If you see any pink areas, stop immediately, as it usually means that you're approaching live tissue that will bleed and cause your goats pain if you cut it. Sometimes, even if there was no blood, you might have still trimmed a bit too far, which can cause some discomfort when walking on certain surfaces.

But what if the hoof is overgrown and folded over the sole? In this case, you should start trimming at the outer side of the hoof, trimming the flap away. Work around the hoof wall until you reach the back of your goat's foot. Sometimes, you may need to trim the heel, but be careful, as this is a very sensitive area.

When you've leveled the hoof, take a look at the dew claws. These are the claws protruding from above the hoof on the back of your goat's leg. They usually require minimal trimming if they are maintained regularly. However, the older a goat gets, the more often you'll have to trim their dew claws.

Once you've trimmed all four feet, let your goat walk freely and watch them walk. Pay attention to their motion, they should be walking normally. This way you can determine if your trimming needs some adjustment for their comfort.

TATTOOING

All dairy goats must have a tattoo before they can be registered with the American Dairy Goat Association. Every goat, except LaMancha goats, should be tattooed in the ears. The LaMancha goats should be tattooed in the tail web area.

The tattoo should be up to 4 letters or numbers, but you can't have one letter followed by a number or numbers. You can't have K2 or D347, for example. You need to tattoo according to your membership identification number, and you can't use a tattoo that someone else has been given. The tattoo should be in the right ear, right tail, or center tail. You should tattoo your goats before they are sold or purchased. In the goats' left ears, you can use a letter that represents the year of birth, i.e., L = 2019; M = 2020; N = 2021 and then, depending on what kid is born into the herd, that can be given a number, i.e., N1, N2, N3, etc.

When tattooing a goat, it should have a muzzle or halter on it. The area to be tattooed should be cleaned with alcohol. You need to make sure the right symbols are set up, then press the thin rubber sponge pad down firmly over the needles. You can check the symbols are correct using paper, then you smear ink on the skin and make sure you are piercing in between veins and cartilage. Green ink is best to use for permanence. Make the imprint quickly and firmly, rub more into it. You could use an old toothbrush to work the ink in, then let the area heal, which usually takes around 1–3 weeks.

Ensure that you keep track of the tattoo numbers in a record. If you are struggling to read the tattoo of a dark eared animal, you can use a flashlight on the outside of the ear.

SHEARING

If you have goats specifically for their fiber, such as Angora or Pygora goats, then they will need shearing twice a year. If the goats are only sheared annually, it can mean that the fiber gets matted in the spring. You can learn to shear the goats yourself, but it is really time consuming and can make your back ache. Instead, you could hire a professional shearer, and they will do this much more quickly and efficiently. Experienced shearers can do around 14 goats in one afternoon.

If your goat has long hair, by having them clipped or sheared, it can make them feel more comfortable. Longer hair is fine in the winter to keep them warm, but not so comfortable in the warmer weather. You may want to ensure that in the winter their coat isn't so long that it gets mud clods attached.

LEASH TRAINING

It is best to start leash training your goats when they are young. If you're constantly around them, be it handling goats, feeding them, or working around them, they will naturally see you as the leader of the herd.

Put a collar and a lead or halter on the goat you want to train. You can also use a dog harness; it is the easiest and safest solution, as it won't choke your goats like a collar can. Start out by introducing them to the idea of a halter or harness. Put it on them when you're around; let them sniff it and play with it. Don't leave it on when goats are not under your supervision, however, as it can catch on to things and goats can hurt themselves.

When they get used to wearing a halter or harness, start in a smaller area and walk forward a few steps. If the goat follows, continue walking. Stop every few steps and reward your goat with a treat. Gradually increase the distance between stops. Lead the goat, and do not let it lead you. If they try to stray away, say "stop" or "get back." Do not drag the goat and tug on the collar. It can block the windpipe and cause the goat to collapse. They may drop to their knees, but they will recover quickly. Never drag your goats, but gently lead them.

Once they feel more comfortable wearing a halter or a vest, you can take them out into a larger area. Remember to take the halter or vest off when you finish training your goats.

CHAPTER SUMMARY

You have goats for milk, fiber, and meat. You have everything down to a fine art now regarding the goats' schedule to ensure they don't have lice and have been dewormed. You groom them with brushes and combs each week, and while you're doing that, you can check the goats' skin and ensure all is well. You trim their hooves every 4 weeks and check them for stones and hoof rot. You have your goats tattooed. You shear your Angora and Cashmere goats twice yearly, and you're able to make beautiful products out of the soft wool from them.

The key takeaways from this chapter are:
1. Be calm and gentle around goats.
2. You will need to use products to prevent lice and to deworm your goats.
3. There are different combs and brushes required to groom a goat.
4. It can be a good idea to put together a goat first aid kit.
5. You need to trim goats' hooves every 4–6 weeks and check for hoof rot.
6. If you want to register a dairy goat, it will need a tattoo that matches your membership number, most goats are tattooed in their right ear, except LaMancha goats, who have their tail area tattooed.
7. If you have goats for fiber, they should be sheared twice a year.

CHAPTER EIGHT

Goat Breeding and Basic Husbandry

It is incredible to see new kids come into the world and fill your smallholding or farm with their energetic cuteness. There is, however, a lot that needs to be done in preparation to breeding, and therefore this chapter will cover everything from preparing to breed to caring for pregnant goats, from birth to newborn checks, and even raising kids. Breeding seasons (known as rut) typically last from the end of July until the end of January. Most goats are seasonal breeders, and they start breeding when there is decreased daylight. It is possible to alter the breeding season by keeping animals indoors and using artificial lighting, but my personal preference with our goats is to keep things as natural as possible. Female goats, which are termed does or nannies, usually give birth to one or two baby goats (known as kids). The gestation period (the length of time a goat is pregnant for) is usually 150–180 days.

PREPARING FOR BREEDING

Does need to keep a good body condition throughout the breeding season, and they need to have excellent nutrition so that they produce healthy babies. If you are planning to produce meat from your goats, then it can be important to maintain a good body condition in the bucks, too. It's worth paying more for a good buck—the best that you can afford. It is possible to use a teaser buck on does. A teaser buck has been castrated and won't be able to make the does pregnant, but they can make the does go into heat. Some signs that a doe is in heat can include:

- Seeking out the buck and paying attention to him
- Eating less
- Behaving in a restless manner
- Peeing more frequently
- Making more vocal noises
- Mounting females
- A swollen vulva with mucus

Although goats can come into puberty as early as 4 months old, it is not advisable to breed them this young. As a rule, a goat should not be mated until it is around a year old. Does may be mated when 10 to 15 months old so that they kid at the age of 15 to 20 months.

Preparing for breeding can be done in a number of ways. It is possible to hand-select the doe and buck, but this can be quite time consuming. Another option is to place a buck into a pen full of does and allow the buck to work out which does are in heat. This is less time consuming initially, but it can be hard to determine the exact breeding record, and you may need ultrasounds to work out when does are expecting. At 1 year of age, the buck should service no more than 10 does at a time (in one month). When he is 2 years old, he should be able to service 25 does at a time. At the age of 3 and older, he can breed up to 40 does at one time, as long as his health and nutritional needs are met.

Another option for breeding is Artificial Insemination (AI), but this can be costly, and it can be tricky to find people with AI experience. Sometimes it can be a farm manager who takes on this role. It is also possible to use out-of-season breeding as previously mentioned using artificial lighting with goats that are housed indoors. This can mean that 60% more of the does can be used for breeding out of season.

It is worth keeping good reproductive records for your goats. This will be helpful if does do not birth on the correct date after being identified as pregnant, or if there are abortions or stillborns. These are things that can indicate bigger issues.

In goats, Chlamydia infection can cause abortions. Chlamydiosis, commonly known as Chlamydophila abortus, is a disease that has no symptoms and stays undiagnosed in a herd until there are many abortions. Does do not have a general pre-breeding Chlamydia screening technique, but bucks' sperm can be examined. It is transmitted through the reproductive fluids of infected animals, aborted tissue, and carrier animals born to infected animals. Pasture and bedding can also get polluted and stay that way for weeks to months, depending on the environment. Chlamydia is identified in goats by examining placental tissue in a lab.

Blood tests are unreliable unless they are done right after the abortion. Chlamydia can be treated with Tetracycline or Tylosin or other effective antibiotics. You should speak with your veterinarian about the best course of action

Practicing proper hygiene and implementing an effective vaccination package are examples of control methods. Buy replacement does and kids from reputable sources with no history of the condition. Be aware that Chlamydia is contagious to humans.

It can be a good idea to increase the amount of food that is offered to your breeding does one or two months before breeding; this is known as "flushing." I recommend that you give the does one-half pound of grain per head every day to improve their condition and help them to ovulate. Ensure the doe isn't overweight, though, because this can cause them to have an uncomfortable pregnancy. If your goat is a milk goat, it can be a good idea to have a little bit more weight on them because once they start milking, it can be hard to get them to increase their weight. Breeding goats is absolutely essential if you have goats mainly for milk. In order to give milk, a goat must first get pregnant and have kids of her own. So, when the does aren't pregnant (called "open"), this can mean that you have a goat without any financial return. We milk the goats for ten months and then dry them off for two months before they have kids again around their next birthday. This dry period is necessary for them to have enough energy to grow their kids.

Two to three weeks before you start the breeding season, the does should be dewormed, vaccinated for tetanus, have their hooves trimmed if needed, and given a vitamin E injection to help with ovulation. If you are planning to use Artificial Insemination, then a cycle should be planned. Trimming the hooves can be important to ensure their feet grow correctly, and because they'll be heavier when they're pregnant with kids, too, putting more pressure on their feet. Increased growth hormone production during pregnancy can make their hooves grow faster, too.

If you're placing a buck with does, you could place him with the does for 42-45 days and then remove him. If you have used Artificial Insemination and it hasn't gotten does pregnant, you could then opt to put a buck in with the does who didn't become pregnant from AI. It can be a good idea to get the buck to wear a harness that contains colored crayon, because this will then allow you to know which does were bred from this buck.

It is always important for goats to have shelter to keep them out of wind, rain, drafts, and sun or heat, but this is even more important when a goat is pregnant for five months. A good piece of advice is that you should spend quite a lot of time handling your goats prior to getting them to breed because if you haven't, they can be tough to work with when they have pregnancy hormones. They're trying to care and protect their kids, so putting the time in to shape their behavior about being handled prior to this is a good use of time.

CARING FOR PREGNANT GOATS

Keep your does calm and as stress-free as possible in the three weeks after breeding, because this is when the pregnancy is starting. Try not to change her daily routine, and try not to travel with her.

The gestation period for a goat is 150 days. It is sensible to perform an ultrasound on goats around the 32nd day of pregnancy. Some goats will have spontaneous abortions. You can also confirm your doe's pregnancy with a blood test sent to a lab at 30 days, and if you chose to, you could run a check for CAE at the same time, which is mainly passed through the mother's milk. It's good to check your goats are clear of it and won't pass it onto their kids.

If your pregnant goat is a milk goat, you can't milk her in the last two months of her pregnancy; she should be dry because she needs to put all her energy towards creating healthy kids. It can be a time to treat does with an intramammary infusion to prevent mastitis, but do remember that there are milk and meat withdrawal times after using this medication.

In the last six weeks of pregnancy, give a booster vaccination for the does of at least Clostridium, C&D, and tetanus so that the antibodies are passed on to the kids and will boost their immunity. Does can have increased feed if they're not up to the body condition they should be. It's recommended in the last 5 weeks for meat goats and 3 weeks for dairy goats.

Goats' body condition is measured on a scale from 1 to 9, where 1 means emaciated and 9 is obese. 70% of the kid's fetus growth occurs in the last 6 weeks, so it is important. It's vital for pregnant does to have high quality hay or alfalfa, and free choice minerals. You can give a second Vitamin E injection to help the embryo develop. Really gently handle does at this time so that they are not stressed.

If a doe is in a good body condition before giving birth, she should produce healthy kids that grow well and have good stamina. You can introduce grain slowly to the doe one month before she gives birth because she will need those extra calories for milk production. Don't add the grain too fast, though, as you don't want to cause bloat.
You can also trim any long hairs around the doe's tail and back of legs about a month before she gives birth; this will make cleaning up the doe after birth easier. Between 8-12 hours prior to giving birth to a kid, the doe's udders will develop, and the area around her vulva will become looser. The doe will lay down and stand up quite a few times. Does should have a calm, clean and dry place in which to give birth.

BIRTHING KIDS
Ensure that the doe is giving birth in a clean, comfortable, and hygienic environment. Fresh straw bedding can be better than wood chips for birth because woodchips can be inhaled and stick to wet newborn kids. Having a clean and hygienic stall will minimize infection.

Have your pregnant doe in a private stall, which will be less stressful and less chaotic than in a herd environment. You can have pregnant does in their own stalls next to one another for company, or you may provide the mother a goat companion.

Make sure that you have a vet's number at hand, or at the very least, someone who has experience in birthing goats. Have some powdered frozen CAE-negative colostrum to feed the kids and disinfect feeding bottles. Have heating lamps, towels, water, disinfectant soap, lubricants, iodine navel dip, gloves, tube feeder, injectable vitamin E, scissors, needles, and syringes at hand.

The doe should be mostly left alone throughout giving birth, and giving birth to a kid should happen within 2 hours after the water sac has appeared. Kids will come out headfirst, leading with their front feet. Occasionally, if it is a more difficult birth, then they can lead with their hind legs. Do be ready to assist with the kidding; ensure your nails are trimmed and clean and use gloves. If your doe is struggling to give birth, then it is advisable to call your vet.

Sometimes, your goats may need cesarean sections; it's good to be fully briefed on goat labor so that you know what to look for. If you need an emergency visit to the vets with a goat, it is likely to cost over $100. If a goat needs a cesarean, it will be over $500, and if you need a vet to come out to the farm for an out of hours visit, this is likely to cost you $800 or more.

NEWBORN CHECK AND CARE

Ensure that newborn kids are breathing and remove any material from their mouth or nose with a clean towel. Keep the newborns warm as well as dry; if bedding becomes wet, change it and put fresh bedding down. If the weather is cold, you can wrap the kids in a towel and use a heating lamp to warm up the temperature. (Nonetheless, make sure this is safely installed so as to not cause injuries, and ensure that the cord cannot be chewed).

The kids' umbilical cord should be cut at 1.5 inches, and then sprayed with 7% iodine to reduce infection. The kids should be kept with the does unless they are CAE positive or if you are bottle feeding them.
The newborn kids should stand up very quickly after being born. Ideally, kids should be given a good amount of colostrum as soon as possible after birth, definitely within the first eight hours. If the kids are weak, help them to stand and suckle. Bottle feed kids with colostrum if the does don't produce enough milk or if they can't stand to suckle. Colostrum can be warmed to 100-102°F (38-39°C) before feeding. You want to ensure kids have had enough milk without forcing them to drink, as that can cause diarrhea.

In my experience, it is definitely worth looking out for newborn kids and their mothers to support them when needed. Here are the things you should look out for:

- Adverse weather conditions of cold/rain—kids can lose energy and die.
- Kids being abandoned by their mothers.
- Sickly does who don't pay attention to or can't provide nutrition for the kids.
- In case of a multiple birth, if a doe has three or more kids, she may not be able to care for and nourish all of them. What can be done is that, if another goat has a single birth, you could wash a goat from the multiple birth in the single goat's amniotic fluid and give it to the single goat's mum, so she thinks she's had two kids. She will usually clean the newborn and care for it as her own.
- Weak newborns who are unable to stand or suckle.
- Predators: keep the area around the birthing stalls and newborns highly protected to avoid any newborns being killed or injured.

- Ensure that none of the herd are being aggressive towards the newborn kids. Sometimes, if a kid got into another pen by accident, the doe in there may try to butt it.

If the kids are raised with the mother, then they will feed off her milk a few times a day. If they are being weaned, then the goat kid should be fed approximately four times a day for the first few days, and then you can move to feeding them twice daily.

RAISING KIDS

Does will usually take good care of their kids, and you will need to give them minimal attention. Does will lick their kids clean after birth and may bleat to the kids to attract their attention. Kids will get up and walk around within a few minutes of being born and should suckle about 30 mins or so after birth. Does will protect their kids from other goats in the herd and will feed and nourish their offspring to help them grow and develop.

By 3-4 months, the kid will be weaned (no longer needing to nurse from its mother). Kids can start to be weaned naturally or off milk when a kid has reached between six to eight weeks, and this will depend on how much forage and water the kids eat. You can offer forage from 2 weeks onwards.

As mentioned previously, goats can become fertile by the time they have reached 4 months old, so it can be important to separate does and bucks when they are young. When they have reached 70% of their adult **body** weight, they can breed, but if you breed too early, it can mean they don't produce as much milk or live as long. It's not a good idea to overfeed them to try to increase their body weight, either. My advice here again would be, for the best health, let things be as natural as possible, and don't force it.

CHAPTER SUMMARY

When we first bred our does, it was wonderful to see the tiny little kids being born and the care their mothers took of them. It's an amazing process, from start to finish. We did have a scenario where one of our does had three kids, and we decided to bathe one of her does in another goat's amniotic fluid who had just had one kid. We placed the goat in a bucket with the fluid and ensured it was covered, then placed it with the singleton pregnancy—and she took to the kid, just as though it were hers, and the kid started to suckle from her. The other doe had two other kids, which were more than enough for her to deal with and to provide nutrition for. All the goats remain in the herd and seem very happy. Goats are enormous fun, and goat kids are just hilarious and so cute!

The key takeaways from this chapter are:

1. Before breeding, handle your doe regularly so that she is used to this.

2. Trim your doe's hooves during pregnancy.

3. Ensure that your doe is a healthy weight and give her high quality hay/alfalfa. In the last month of pregnancy, you can slowly add some grain. Don't forget the minerals, too!

4. You can confirm pregnancy with an ultrasound or blood test.

5. Do a CAE test when pregnant, so that mothers can't pass this onto kids through milk.

6. Ensure the doe is dry (not being milked) for the last two months of pregnancy.

7. Make sure that, throughout pregnancy, your doe has good shelter with no drafts, and before birth, that she has her own individual pen with goat friends nearby for company.

8. Make sure pregnant goats have had the CD&T vaccine so that they will pass on immunity to their kids.

9. Have a clean stall for the doe to give birth in.

10. Have plenty of powdered or frozen colostrum.

11. Have contact numbers of vets or goat mentors.

12. Watch for changes in goat behavior that show birth is imminent.

CHAPTER NINE

Milking Your Goats and Processing the Milk

HOW TO MILK A GOAT

Before you begin, make sure you have everything ready that you need for milking and straining. The straining cloth should be boiled, the jars and funnel sterilized—see the next section for information about doing this. Fill a food bowl with your treat feed and place it in the milking stand's feed bucket area.

The amount of treat feed to give a goat will depend on how much milk she gives. Some goats produce more milk with higher amounts of grain, while other goats seem to do better with less grain. Approximately two cups of grain is a good amount to start with. Have your milking bucket close by, but not anywhere that the goat can easily knock over. I keep a small table near the milking stand for my bucket and jars.

Walk the goat up to the milking stand, guide her head through the headgate, and secure it around her neck. You will now need to clean her udder. Either brush it with a dry cloth or the back of your hand to remove stray hairs and dirt, or if she's very messy, you can wash her udder with a wet cloth. Rub it gently with a very dry towel, making sure it is dry, as you're far more likely to get sick from dirty water dripping into the milking bucket than you are from a few stray hairs or bits of dirt. If you're washing and drying the udder, you will need to use a separate cloth for each goat. Few goat books recommend following the first method (my preferred one) for udder cleaning and prefer the more thorough washing and drying approach. A lot of the belief in washing goat udders comes from milking cows, as they seem to be attracted to the muddiest part of the paddock and can have very dirty udders. On the other hand, goats will find the driest place possible and don't seem to get dirty often. The simple method works for my family, as we keep our goats clean and dry with lots of straw, but if your goats are covered in muck, washing and drying is the best option. My mention of the simple method might be a controversial approach, but it is far easier to brush the udder quickly than to wash and dry the udder. If you're in doubt, can't chill the milk quickly, or are sensitive to food contamination, wash and dry the udder, making sure you dry it thoroughly.

Make sure you are seated comfortably. I sit on the edge of the milking stand, but plenty of people use stools instead. You should be able to sit there milking the goat with no need to bend your back and without stretching your arms out awkwardly to reach the teats. Take two squirts of milk from each teat and milk it onto the milking stand, ground, or a separate dish. By discarding the first squirts of milk from each teat, you decrease the risk of contamination from anything lurking on them. Once you have discarded these first squirts and your goat's udder is clean, place the milking bucket close to the udder and begin milking into it.

To milk a full-sized goat, first place your thumb and index finger around the top of the teat where it meets the udder and close it off, then close your middle and ring fingers (or just the middle finger if her teats are small) around the teat to squeeze the milk out of it. Repeat this with one hand after the other until it becomes more difficult to get the milk out of the teat. Remove the bucket, and then massage her udder or mimic the action that a kid uses on it by pushing against it with your hand, then place the bucket back under her and continue milking as you were before. This helps her to let down as much milk as possible.

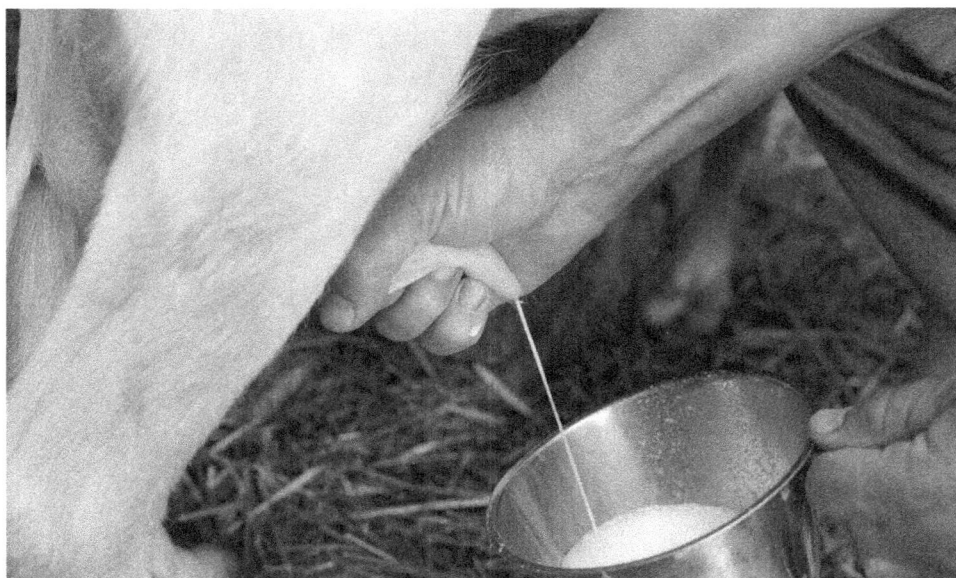

If your hands get tired using that method of milking, you can alternate with another method where the thumb is not used, and just the index finger is used to close off the top of the udder. This method uses different muscles to the usual milking method but can't usually be done until the udder has emptied a bit. To get the last of the milk, "strip" the udder using both hands on one half of it at a time. Gently squeeze the milk from that half of the udder into the teat and then out into the bucket. When it's time to stop milking, she will be giving the tiniest bit of milk (or none at all).

It's easiest to watch someone else milking to learn, and if you don't have anyone nearby, try searching for videos online. To get your hands used to the action of milking, go through the motions of milking on your thumb. It's a good idea to learn to milk while the kids are still drinking their mother's milk because the kids can help drink the rest of the milk, and your doe is not at risk of udder problems or drying up as long as someone is taking the milk

Once you are in a good milking routine with your goat, milking will take less time than straining and cleaning. I take around five minutes to milk a goat with a good udder—a bit longer if her milk is slower to flow. When you've finished milking your goat, leave her on the stand to finish her meal while you strain the milk. This extra time on the milking stand helps the teat close before any bad bacteria can get into it. To strain the milk, place a funnel over the top of a glass jar, then cover the funnel with a sterilized thin cloth, such as butter muslin or cheesecloth, and pour in enough milk to fill the funnel. If it's taking a long time to go through the cloth, you may need to find a thinner or more loosely woven cloth next time. Alternatively, gather the edges of the cloth in your hands and tilt them around carefully to get more milk through. The creamier the milk, the longer it will take to strain.

A MILKING ROUTINE

1. Boil the straining cloth—you can do this the night before.
2. Assemble everything you need during and after milking (e.g., clean jars and funnel on the table, clean milk bucket near the milking stand).
3. Bring the goat to the stand, clean her udder, and milk her.

4.   Strain the milk.
5.   Take the goat back to her paddock.
6.   Repeat for other goats.
7.   When you've finished milking all the goats, wash the straining cloth and hang it up to dry. Sterilize the bucket and funnel.

MILKING EQUIPMENT AND HYGIENE

When the milk is in the udder, it is sterile and is a perfect food for baby goats to drink. You rarely hear of baby goats getting sick from their mother's raw milk in the same way that the media sensationalizes cases of humans getting sick after drinking raw milk.

Due to its neutral acidity, any milk out of the udder, raw or pasteurized, is an easy medium for bacteria to grow in. This can be beneficial when we encourage good bacteria to make cheese, yogurt, and other fermented foods, or it can be bad. If you're drinking the milk right away, hygiene is not much of an issue. The longer you wish to store it, the more careful you need to be about the storage conditions and the likelihood of anything coming into contact with the raw milk.

There are two main approaches to preventing contamination by the wrong bacteria on milking equipment. The most sustainable and healthy option is heat sterilization. Any containers used for the milk must be heated to a safe temperature first, then the milk can be added. The other main method is with chemicals.

After the equipment has been sterilized, it needs to be kept away from anything that could contaminate it. I recommend milking buckets with lids—that way, you can keep the funnel and the bucket's interior away from potential contaminants. I use a 7 liter bucket, but around half this size is sufficient. Empty the bucket as soon as you've milked each goat to avoid it being kicked or stepped on, losing the milk. Buckets larger than 7 liters will be difficult to fit underneath the goat, so search for a small stainless-steel bucket with a lid. It's tricky to find, but it's something that only needs to be purchased once, so it's worthwhile getting the right one to start with. A stainless-steel stockpot can be used instead of a bucket. Try to find one without air holes in the lid, and you will have something that works just as well as a bucket with a lid.

Heat sterilization can be done with boiling water or in the oven. Using boiling water can cause injuries (and rude words), so the oven is often preferable. Everything you put in the oven for sterilization should be metal or glass—never plastic, unless it is rated to withstand high temperatures. You can use the oven to quickly dry your plastics after safely boiling them, but sterilization in the oven takes time. Anything put in the oven for long enough to sterilize it should be able to withstand the heat. I recommend stainless steel milking buckets, stainless steel funnels, and glass jars.

Oven sterilization: To sterilize using the oven, place your clean bucket and jars in it upside down and switch it on to 230°F (110°C). Don't use ovens with visible (orange-glowing) electric elements for sterilizing glass, as the glass can crack. Leave the oven to heat up, and once it's been fully hot for at least five or ten minutes, test it by touching something sterilizing in there it should be very hot to the touch. Turn it off, close the door, and leave it to cool down.

Boiling water sterilization: First, jar lids and plastic funnels should be submerged in boiling water for 30 seconds, and then the water drained off. Place them in the oven as it cools down to dry them and keep them sterile until you need to use them. Be careful that your plastic will stand up to this, as some plastics are flimsier than others, but the plastic Ball mason jar lids will without any problem, as do good funnels designed for jam.

For the straining cloth, I rinse it and hang it out to dry after every milking, and then boil it when we're ready to begin milking the next time. If you need to sterilize everything with boiling water, boil enough so you have plenty to pour over everything that needs to be sterilized. The aim is to heat the jar, funnel, and bucket surfaces to kill off any potential nasties that might be lurking there, so slowly pour plenty of boiling water over all the surfaces.

The jars should be wet (but with no standing water) before adding boiling water so they don't break. Pour the water into the jars, filling them to around one-third of the way, and put the lids on. Turn the jars on their sides and rotate until the water heats the jars and you can't comfortably handle them. You can rely on a wood cooking stove that's often slow to boil water in the mornings, so at night you put the straining cloth in the bucket with the funnel and leave it all submerged in the boiling water overnight, draining it all in the morning.

Every time you empty a jar of milk, rinse it twice with a little cold or lukewarm water and leave it until you're next ready to sterilize jars. Leaving jars sitting with a little milk in them makes it hard to remove it later.

PROCESSING, HANDLING, AND STORING THE MILK

Once the milk is in the jars (if you're not drinking it right away), you should chill it quickly. I surround each jar of milk with ice-filled containers—I use the "ice bricks" made for coolers. You can even chill the jars before you milk, so the cold jar begins chilling the milk as soon as it is strained.

After the milk has been chilled, it's important to keep it at a low temperature and not to let it fluctuate. Many people keep their milk in the refrigerator door, but this is not a good place for longer-term milk storage. Only store it there if you will be using it in the next 24 hours. The temperature fluctuates too much in the fridge door, and it's also the warmest part of the fridge. It is best if the milk is on a shelf in the fridge, preferably toward the back. Avoid leaving the milk sitting out at room temperature for entire mealtimes or similar periods unless you plan on drinking the rest of the jar soon.

Don't place warm containers of leftovers or anything else next to the milk. If you're adding a lot of room temperature and warmer things to the fridge at once, add ice-filled containers right next to the milk to make sure that it stays cold.

The quicker you'll be drinking the milk, the more relaxed you can be about its storage. Bacteria need both the right temperature and the right amount of time to multiply to dangerous levels.

Now that I rely on a small off-grid system without a fridge, I find that in winter, we drink all the milk within around 24 hours, so the temperature of an unheated room out of the sun (50°F/10°C) is cold enough for it, even without ice bricks. In summer, there is a lot more milk to handle. On the hottest days, anything that doesn't get purposely fermented into cheese or yogurt begins to ferment on its own within 24 hours unless I'm super careful about switching the ice bricks around.

1-quart (1-liter) mason jars and 22 ounces (650 ml) Passata bottles are good sizes for storing fresh goats' milk. To make the easiest "chèvre" from milk warm from the udder, it is worth having a 2-quart (2-liter) jar for this purpose.

CHAPTER TEN:
Harvesting Goats' Meat

Many people in the US have been raised to eat chicken, pork and beef, but goats may not immediately jump to their mind when planning dinner. But many countries have goat as part of their menu for a good reason. Goats are versatile and easy to care for, requiring less land than cows. Goat meat has less fat than other types of red meat, and the fact that it's so lean can make it a popular choice among consumers.

If you want to store meat from an animal, it helps to know that cows produce a lot of meat, whereas the meat from a goat is much more manageable to store. A key tip for people raising goats in America for meat, prior to purchasing them, is to check the zoning requirements, because in different states goat definitions vary from "livestock" to "companion animals," and this could have an impact on how successful your business is.

Some breeds of goats bleat more loudly than others, so if you are in a suburban area where neighbors could complain about the noise, this is worth thinking about. You need to consider whether there is a market to support the number of goats you have. Do you want to raise goats for meat for yourself or to sell? Will you be able to sell all culled animals? Will you breed goats to make replacement goats or buy new ones? You need to think about this before purchasing goats. So, assuming you have done all the above research and have your goats, you may wonder at what stage of the goat's life you should butcher it for meat. The answers are below.

WHEN TO BUTCHER A GOAT

It is generally bucks that are harvested because does tend to be kept for breeding and producing milk. A good age for a goat to be butchered is 8-10 months old. It's not just about when a goat is this high or weighs that many pounds but about how much the goat will yield. This will vary from breed to breed.

To evaluate when your goat is ready, you will need to look at how much muscle it has on its forearm and its overall appearance. The more muscle your goat has, the more meat you will get from it. A goat's forearm is the top of its front leg where the leg meets its body. Your goat will grow taller and bigger first and then will develop muscle after that. If your goat doesn't have much muscle, allow it more time to grow.

Another place that you can check the muscle of a goat is the outward facing muscle on the goat's right rear leg where the leg joins the body. This area is called the stifle, and this part of the goat should have a definite outward bulge when you are viewing the goat from behind. The goat should have muscle in its inner thighs that joins the center of its body. If the goat doesn't have muscle, leave it longer to grow and develop.

If you still have the doe that gave birth to the kid, compare the kid with her; the kid should have more meat on it than the doe. You can also take pictures of the kids from time to time, at the front, rear, and side, to give yourself something to compare its growth to. It can be easier to see this in photos because when you're looking at the kids twice a day at feeding time, you don't tend to notice gradual changes.

As a general rule of thumb, whatever the carcass of the goat is, you will yield about 50% of that in meat. So, if your goat weighs 80lbs, you'll get approximately 40lbs of meat. The goat you butcher should be approximately 8-10 months old. Sometimes people want a 50-60lb roaster goat, which is younger than 8 months.

If you want to get your goat to a point where it can be butchered sooner, you can feed it some grain or pellet feed. Generally, goats that just have forage take longer to reach a weight where they can be butchered. If goats have been fed a concentrated diet, less off-flavors are likely to be reported, compared to goats that were fed from the range. Boers are a meat goat, and they will need supplemental food in order to grow to an optimal weight to be butchered. If you just want your goats to only forage, then a better breed to choose is Kiko or Spanish.
Many people try to get a goat between 80-100lbs in weight. You can weigh your goat each month to see how they are progressing. If you don't have a livestock scale, you can lift the goat onto your normal bathroom scales in your arms, then deduct your own weight to work out what the goat weighs.

If your goat isn't looking as well as you'd like it to, you need to ensure it has been dewormed and that it has more calories to eat. It may also simply need more time to develop muscles. You can't hurry this along too much—goats need time to grow.

HOW TO BUTCHER A GOAT

To butcher a goat, it is a similar technique to butchering a lamb or deer. Simply put, you'll gut the animal, remove the head, work the skin off, and rinse the carcass once this has been done. If you are planning on butchering a goat, you can do this when a goat is approximately 8-10 months old. If you are butchering the goat yourself, you will need to slit its throat, but it is kinder to make the goat unconscious first.

A goat's head is very hard, so you will need to be strong and use a very heavy implement. When the goat is stunned, you will be able to string it up, which will make the next stages easier. If you save the blood from the goat, it's recommended that you add a splash of red wine to it and stir it thoroughly; this will prevent the blood from clotting into a lump.

You can use goat's blood in recipes—for example, blood sausages, but you can also use it in rice dishes, too. Some people, including many top chefs and restaurateurs, believe that if an animal is killed for food, it makes sense (and honors the animal and its sacrifice of life) to use as much of that animal in every way you can.

In order to skin the goat, start by cutting the skin at the back legs, then move downwards to separate the skin from the body of the goat. There will be a membrane attached to it, which can be loosened with a knife. You can cut around the goat's anus and remove the genitals. The goat's testicles, otherwise known as "sweet meat" or "sweetbreads," are a delicacy that can be poached and then cooked in something like an omelet.

If all the skin has been loosened with a sharp knife, you should be able to pull it downwards to remove it. You do need to be strong and firm when you are pulling. If you start to tear the meat, you need to loosen the skin some more with the knife. When you have done this, you can cut the goat's head off with a sharp butcher's axe. You can cook the goat's brains if you wish to do so, and the skull can be split between the horns using the butcher's axe.

You need to gut the goat next, so start by cutting between the legs, down towards the rib cage, and have a bowl to catch all the innards that you pull out. You can tie the intestines so that they don't leak over your meat. You can put the liver and kidneys to one side to be used for cooking. You can tear the diaphragm area to remove the heart and lungs and put them with the liver and kidneys you have set aside. You will want to remove the intestine and any other tubes or bits of gristle. You can tie the top of the intestines to pull out. This will take a lot of work and strength to pull out. You need to chop off the feet using the butcher's axe, and then the goat carcass needs to be given a really good wash. It is best to hang the carcass after skinning for 24 hours before you cut that up into relevant pieces. If you are trying to utilize as much of the animal as you can, you can boil the bones to make soup.

You will want to remove tenderloins first, using a knife to start where the loins rest behind the ribs. Then remove the rear legs; you can use a hacksaw to cut through the spine. You next want the flank steaks, then the back strap, before getting the rib meat from the carcass. Then next is the shanks from the front legs and the shoulders, and finally the neck meat can be used for stew.

I would, however, advise that in most instances, you employ an abattoir or a butcher to perform this job for you, rather than doing it yourself—especially if you've never butchered an animal before. It will also allow you to watch and learn, and once you're confident enough, you can try to butcher your goats yourself.

MEAT QUALITY

Research has shown that when people have goat meat from younger animals, the meat is more tender, and doesn't taste "off" compared to older animals. If meat is from a dairy goat or a meat goat it can make a real difference to the quality of meat, too. Cabrito is meat from young milk-fed goats that weigh between 15–25lbs, and they are usually harvested at a month to two months old. People tend to barbecue this type of meat. Chevon is the name for meat coming from goats who are 6–9 months old, they have eaten forage and some grain and typically weigh 50–60lbs. Cabrito meat is more tender than Chevon, it has a similar tenderness to lamb. If a goat is larger than Chevon, this is still meat that can be sold. If goats have increased fat content from high protein grain for a long time, this can increase the aroma and flavor of their meat. But the older a goat is, the tougher the meat will be. Goats can be butchered at up to 16 months of age, but it would be very rare for this to happen. Bucks are usually harvested before they reach puberty because then bucks take on a distinctive smell, which can permeate their meat. Most young bucks are wethers that have been castrated.

Goat meat is lean and bright red and doesn't have the fat marbling that you see in cows. Young goats have thicker muscles than older goats. If you are buying goat meat from a butcher, you want good, thick, meaty shanks, not thin or narrow ones.

STORING GOAT MEAT

If you intend on using the goat meat very soon, it can be refrigerated. If you want it for the longer term, then freezing is a better option. If you have small chunks of Chevon or it's ground meat, you can keep it in your fridge, but use it within two days. If you have whole cuts of meat in the fridge, they are fine there for 3-5 days, but they need using at that point. If goat meat has been properly packaged and frozen, it can be kept in the freezer for approximately 3-4 months; this is for ground or cubed Chevon. Larger cuts of meat can be kept frozen for a bit longer, up to 6 months, such as steaks, chops, or legs. If it is kept frozen for longer than this, it can taste off and be damaged by the freezer. It's good practice to label any freezer meat and date it so that you know exactly how old it is. When frozen goat meat is defrosted, it's best to thaw it in the refrigerator; it's the safest way to do it. Depending on the size of the meat, this can take 24 hours or longer. You want to ensure the goat meat is in a leak-proof container, so that it does not leak any blood elsewhere and contaminate other food in your fridge or create a mess. You can also defrost meat in airtight packaging in cold water or use a microwave, but the fridge is the best option. Never refreeze thawed meat because once the meat reaches a certain temperature, bacteria could multiply and make you sick.

CHAPTER SUMMARY

I still remember the first time I tasted goat meat. To be perfectly honest, I was a bit hesitant to try it initially, because I love goats and it was a different type of meat to the beef, pork, lamb, or chicken that I'd eaten throughout my childhood. It was unknown to me. But it is beautifully tender, delicious, flavorsome, and much healthier than many other types of meat.

Remember:
1.   Goat meat is very lean.
2.   It's relatively easy to store goat meat because they are not huge animals.
3.   It's more typical for bucks to be harvested for meat rather than does.
4.   Typically, a buck should be butchered at 8-10 months.
5.   Check how much muscle the goat has before butchering; if it does not have enough, give it more time to grow.

6.   To butcher a goat, gut the animal, remove the head, work the skin off, and rinse the carcass. Or preferably get a butcher to do this for you.
7.   You will need a butcher's axe to butcher a goat.
8.   Younger goat meat is generally more tender. Cabrito is harvested at 1-2 months old. Chevon is goat meat from a goat 6-9 months old.
9.   Goat meat can be refrigerated but needs eating quickly: 2 days for cubed and ground goat meat and 3-5 days for whole cuts.
10.   It is best to thaw frozen goat meat in a refrigerator.

As well as goats being fantastic for meat and dairy, there are many additional benefits of having goats, weed control, manure, using goats as pack animals, and having them as pets.

CHAPTER ELEVEN:
Dairy and Meat Products

So, you've diligently followed all the steps in the book so far regarding getting goats, housing them, providing adequate fencing, feeding them a nutritious diet, taking good care of their health and well-being, breeding them, milking them, and harvesting goat meat. You may now wonder what you can do with the product of all that labor. This chapter is here to help with details about cheese making, creating other milk products, and some goat meat basics to get you started.

CHEESEMAKING

We tend to drink the freshest goat's milk, and we have a supply that just continuously renews itself. We use day-old milk for cooking and cheesemaking. If you have milked your goat, and you live in the US, you need to decide whether you will sell the cheese you make from it or just have it for your own use. If you decide to sell it, then you need to either pasteurize your milk or let the goat cheese age for 60 days before you can sell it.

Goat cheese, also called Chevre, is super easy to make at home, and on top of that, it's ridiculously cost effective. With just three ingredients, you can make a batch of lovely Chevre for your family and friends to enjoy—or to sell.

Goat cheese is made via acid/heat coagulation, which is the oldest method of cheese making in the world. Essentially, lemon juice or white vinegar break apart the protein structure of the milk once it's reached a certain temperature.

Here are the products you'll need to make a pound of Chevre:

1 quart of fresh goat's milk
1/3 cup of lemon juice or white vinegar
¼ teaspoon non-iodized salt

And the equipment:

A pot to boil milk
A colander to drain the curds
Digital thermometer
2 cheese cloths

Once you've gathered the ingredients and the equipment, it's time to start making cheese. Here are the steps you have to follow to make your own lovely goat cheese:

1. Slowly heat the milk to 180-185°F; use the thermometer to check the temperature of the milk. The surface should look foamy with bubbles forming. Turn off the heat once the milk reaches the desired temperature.

2. Stir in the lemon juice or white vinegar with a long-handled utensil. Let this mixture sit for 10 minutes. The milk will start to curdle and should become thicker on the surface.

3. Line a colander with 2 damp cheese cloths and suspend it over a large pot or bowl. Pour the milk gently into the colander lined with cheese cloths. Cover and let it drain for 2 hours.

4. Gather the cheese cloth up around the curds and tie it into a bundle. You can use a rubber band to tie it up. Hang the bundle over a pot so that the liquid can drain. You can do it by tying the bundle to a long utensil and putting it over a pot. Let it drain for 4-8 hours. The longer you let it drain, the thicker the cheese will be.

5. Once the liquid is drained, untie the bundle and put the cheese into a bowl. Mix in the salt to taste.

6. Form the cheese into a log or a wheel with your hands, just like you would form batter. You can use a cookie cutter to mold cheese, as well. You can use the cheese right away or refrigerate it for up to 2 weeks. Enjoy!

MAKING YOGURT

A simple recipe for goat's milk yogurt can be found online. Lactose intolerance sufferers may find goat's milk to be a preferable option because of its low lactose content. You can either buy a yogurt maker or utilize a cooler that can hold two 1-quart jars.

What you'll need is the following:

12 cups of plain yogurt with live culture or 1 five-gram packet of freeze-dried yogurt starter

2 quarts pasteurized goat's milk

2 teaspoons gelatin

Candy thermometer

Yogurt maker or 9-quart flip-lid cooler

Method:

1. Pour 14 cups of milk into a bowl and add the gelatin; do not stir, simply allow it to settle on the surface. Put away.

2. Pour the remaining milk into a pan and place the candy thermometer inside. Heat the milk to 180 degrees Fahrenheit, then reduce the heat to low and stir for 20 minutes.

3. When the milk reaches 120°F, add the gelatin mixture from step 1. When the milk reaches 108°F, place it in an ice bath. Add the yogurt starter culture or powdered yogurt culture. Mix thoroughly but gently.

4. Fill the cooler halfway with boiling water. Put the cultured milk into two 1-quart jars, seal the lids, and place them in the water. Close the cooler or yogurt maker and allow it to incubate for 5-10 hours. The longer it is left to incubate, the firmer the yogurt becomes, and the tarter.

5. Refrigerate the yogurt until you use it; only keep it refrigerated for up to a fortnight, then discard.

MAKING KEFIR

Kefir is a lovely and refreshing fermented milk drink. Everyone in my family loves kefir. I used to drink it since I was a little kid, and I absolutely love it to this day.

To make kefir, you need milk kefir grains or a kefir starter culture. Using a direct-set powdered kefir starter culture is a good option to make kefir without milk grains. All you need is a start culture and a quart of pasteurized goat's milk.

Pour the milk into a glass or plastic container and add the kefir starter culture until it has dissolved. Cover the container and place in a warm spot at 72-74°F for 12-18 hours. Once it has finished, put a lid on it and keep it in the refrigerator. I prefer to drink it plain, but you can flavor your kefir with fruit, honey, spices, or sugar if you wish.

MAKING KUMIS

Kumis is a healthy and refreshing drink made from goat's milk. It has a peculiar sweet taste. It can aid digestion and restores microflora to the upper respiratory tract, which can be a great aid to health.

To make it, you need a quart of the freshest goat's milk, a glass of boiled water, 3 tablespoons of kefir, 3 tablespoons of honey or sugar, and 5 grams of baker's yeast. You pour the boiled water and sugar into the milk, then put a teaspoon of kefir into the mix at a time; it should mix without lumps or flakes. Wrap the container in a cloth, and place in a warm place overnight or for 5-6 hours. Then the thick substance will need filtering through a strainer. Add water to the yeast and mix this into the kumis; leave it for 5-10 minutes. It can then be poured into bottles and refrigerated. After a day of it being in the fridge, you can drink it. It should definitely be drunk within 3-5 days.

MAKING BUTTER FROM GOAT'S MILK

Goat milk butter has a lower melting point than butter made from cow's milk. It contains more unsaturated fatty acids. Goat milk butter is pure white in color. It takes longer for cream to form in goat's milk unless you have a separator (but it's hard to find these, except second-hand).

In order to make goat milk butter, you will need the following equipment:
- Dairy thermometer
- Glass churn or a mixer
- Pan that can fit into a larger pot of water

And these products:
- Goat milk cream
- Cold water
- Salt to taste

First, you need to separate the cream in goat's milk. There are two ways to do it. You can purchase a cream separator, which will make the job quicker and easier, or you can simply put a jar of milk into the fridge and refrigerate it at least overnight (better yet, a few days). The heavy cream will rise to the top as the milk cools down. If you let your milk sit for 3-7 days, you'll get about ½ inch of cream that's risen to the top.

Scoop the cream from the top with a spoon until you get to a more watery part. Try to get as little of the watery milk as possible. I suggest getting a quart of cream. You can store it in the freezer until you've collected a full quart. Don't worry; the cream will stay fresh in the freezer even if it takes you a month to gather as much as you need. Now that you have the cream, it's time to move on to making butter.

Here are the steps you need to follow to make your own lovely goat milk butter:

1. Once you collect the cream, let it stand overnight so that butterfat globules become ripe.

2. Heat the cream in a double boiler to 146°F, put the pan in cold water, and let it cool down to somewhere between 52 and 60°F.

3. Pour the liquid into the churn or mixer (only half full) and start

agitation. If you're using a churn, the butter should come 30 to 40 minutes after this. If you're using a mixer, mix on high for about 10-15 minutes. If the blender starts bogging down, you can add a little goat's milk to make the mass less dense.

4.   Once you can see larger balls of butter forming, stop churning or turn off the mixer and scoop out the balls of butter.

5.   Place the butter into a bowl of cold water and wash out the excess buttermilk by squeezing the butter with a spoon. Repeat the washing process a few times until the water is clear.

6.   Once the butter is clean, squeeze any excess water from it with a spoon

7.   Store the butter in the fridge. Some excess water may still come out of it during the first day; just pour it off.

And that's it. Enjoy your own homemade goat milk butter! If you need it for later, you can freeze the butter for up to 6 months.

BEST GOATS MEAT

Frankly, goats from any number of breeds end up getting slaughtered for their meat. However, not every breed is bred specifically for meat production. The one breed that stands out more than the rest is the Boer, which is a breed best known by the people living in the Upper Midwest of the United States associated with meat production. Developed in South Africa, the Boer is one of the few breeds selectively bred for meat harvesting. Incredibly, the Boer is a newer addition to the United States and was first imported in 1993 from New Zealand.

Other goat breeds that are generally acknowledged as meat producers are Spanish, Pygmy, Kiko, and the Myotonic (fainting goat). Even if you primarily raise dairy goats, sooner or later, most owners will decide to cull the herd's least productive animals. Those targeted will be low milk producers, poor mothers, and the least valuable of the newest bunch of kids. Especially if you are a dairy producer, you may want to consider culling if you have had a bumper crop of male kids because typically, you only keep one out of every 100 bucks in the eventuality that you need to replace an aging patriarch buck.

CURED MEATS, MEAT DRYING, AND SMOKING

Prior to applying any preservation method to goat meat, it must be fresh. Curing should not be used to save meat that has developed an excessive amount of bacteria or has spoiled. Goat meat does not require aging because curing/smoking tenderizes it. If aging is needed, cool all meats to temperatures below 40°F.

Salting

Use only food-grade salt-free additives such as iodine. Using salt tainted with impurities can have unfavorable effects. Thawing must be closely monitored and regulated to ensure accuracy and avoid temperature manipulation. Inadequately thawed meat can result in inadequate cure penetration. Abuse of temperature can result in spoilage or the growth of pathogens.

Compounds for Curing

Purchase ready-made prepared cure mixes and carefully follow the directions on the package, or carefully blend cure mixes at home using an accurate scale.

For dry-cured products that will not be baked, smoked, or refrigerated, use nitrate-containing cure mixtures (e.g., Prague Powder 2, Insta-Cure 2). Dry cure with no more than 3.5 oz. nitrate per 100 lbs. meat or wet cure with no more than 700 ppm nitrates.

For all meats that need frying, smoking, or canning, use cure mixtures containing nitrite (e.g., Prague Powder 1, Insta-Cure 1). Dry cure using no more than 1 oz. nitrite per 100 lbs. beef. 14 oz. per 100 lbs. For sausages, a concentration of 120 parts per million is normally adequate and is the limit allowed in bacon.

Nitrites are harmful when used in excess of recommended amounts; thus, caution should be exercised when storing and using them. A lethal dose of sodium nitrite is approximately 1 g or 14mg/kg body weight in an adult person.

Using sodium nitrite instead of sodium chloride in standard curing recipes may result in a lethal dose of nitrite in the improperly cured product. As a result, purchasing and using curing mixtures instead of pure nitrites is better (saltpeter).

Cure Penetration

Cure mixtures are ineffective at penetrating frozen meat. Prior to curing, it is important to fully defrost meat in the refrigerator. To ensure uniform cure penetration, pieces must be prepared to uniform sizes.

This is important when it comes to dry and immersion curing. Utilize an accepted recipe to calculate the precise amount of curing formulation to use on a given weight of meat or meat mixture.

When using a dry-curing process, all surfaces of meat must be rotated and rubbed at appropriate intervals to ensure cure penetration. Immersion curing involves periodic batch mixing to ensure consistent curing. Curing can take place between 35°F and 40°F.

The lower temperature is chosen to maximize cure penetration, while the upper temperature is chosen to minimize microbial growth. Curing solutions must be discarded unless they are used on the same batch of the product during the curing phase do not reuse brine due to the risk of bacterial growth and cross-contamination.

Smoking

Ascertain that the smokehouses are operating properly (e.g., heat, airflow, moisture). Temperature gauges that are adjusted properly should be used (for cooking temperature and meat internal temperature). Procedures for properly thermally treating cooked meat in accordance with the Food Code must be established and implemented. Without proper cooking, smoke is ineffective as a food preservative. Caution should be exercised while smoking meats at dangerously high temperatures (40-140°F) for extended periods of time. In this case, the meat must have been salted or cured prior to cooking.

CHAPTER TWELVE

Starting Up and Improving Your Business

WHAT ARE THE BENEFITS OF GOAT FARMING?

Goats are one of the earliest domesticated animals owned by humans. They appeared in history books and in the Bible, where they were mentioned in detail. Their uses today remain largely unchanged from when they were first domesticated.

If you live in a rural area, you may want to consider goat keeping. When you visit a farm, you are often greeted by hens, chickens, cows, and horses. These creatures are extremely valuable economically and domestically. However, if you're looking to expand your farm's animal population, you might consider goats.

The following are the most compelling justifications for starting your goat farm.

1. Goat's meat provides just the right amount of protein.

Due to their small scale, goats are a more favored meat source than cows and pigs. In effect, since these animals do not occupy a large amount of space, they are easier to maintain or store. Also, while larger carcasses necessitate more time spent thinking about and implementing preservation techniques, goat meat almost eliminates the need for this. Your family will eat goat meat before the next one is slaughtered.

2. You can make different dairy products from goat milk.

Goat's milk is superior to milk provided by other animals in many ways. According to experts, it is more nutritious and suitable for nursing infants who are allergic to cow's milk. Also, if you know how to process goat's milk into other dairy products like cheese, you can earn a lot of money.

3. The hair of goats may be fashioned into garments.

Since Biblical times, goat's hair has been transformed into exquisite scarves and other articles of clothing. Many people prefer goat hair clothing because it is lightweight and acts as a natural heat insulator. They provide necessary warmth on cold days but maintain a cool temperature during the summer.

4. There are many uses for goatskin/hide.

The goat's skin can be transformed into stunning accessories, belts, and shoes, incorporated into furniture items, or used as wall decor and centerpieces.

Goats have even more uses than these. For example, using organic gardening methods, their manure can be converted into instant fertilizers. You may either use their manure in your garden or profitably sell it, and as herbivores that consume mostly grass, goats can be used to clear forested areas by consuming the weeds effectively.

Goat meat is a healthy alternative. As people become more health-conscious, they make healthier choices. In comparison to red meat from other species, goat meat is leaner and lower in fat. This increases its appeal among health-conscious consumers.

In comparison to other animals, goats are simpler to breed. Also, they are less particular about the foods they consume and are more adept at adjusting to extremes in temperature. This simplifies their rearing and treatment in comparison to other livestock.

In comparison to other animal-based companies, goat farming needs a much smaller initial investment. Also, goats are docile by default, which makes them easier to care for and maintain.

Goats have a four-month gestation period and are prolific breeders. You can start a goat farm with just two goats, and within two years, you are guaranteed to have a couple more goats, resulting in increased profit. Goat milk is better than cow milk because it causes no allergic reactions and is more easily digested.

Therefore, if you want to raise animals for meat or milk, goats are the best option. If you have the proper knowledge, there are a lot of opportunities to earn significant income.

It's important to remember that it's not just about taking care of whatever kind of goat is available; you must choose the proper breed to produce the highest quality meat that you can offer. For instance, you can select specific breeds that have been said to produce the highest quality meats—Spanish goats and Boers, for example. They are considered the best breeds for producing meat because they have larger bodies than other breeds but do not accumulate as much fat.

Also, raising goats for meat is a relatively simple task, particularly because they mature quickly; this means that you can produce meat in a relatively short period. The only thing you need to keep in mind is that even though they produce the most meat, you must take good care of them during the process because they require a special diet.

On the other hand, some believe that different breeds such as fainting goat and brush will yield the most meat. The positive aspect that people notice about these goats is their ability to withstand any form of diet, temperature, or even weather.

Even though the Brush and fainting breeds of goats do not have the same size bodies as the Boers and Spanish goats, people take pride in the fact that these breeds reproduce at a higher rate than other breeds. All experts in raising goats for meat know that this is a bonus because you will still end up with an abundance of meat to offer. Along with selecting the appropriate breed for your goats, you should always have a veterinarian examine them to ensure they are safe.

Goat meat production is the fastest-growing agriculture segment in the United States, accounting for an estimated 70% of all meat globally. Raising meat goats can be quick and straightforward if one follows the advice of those who have done so.

Some states and countries exclude agricultural land from taxation. Farmers are eligible for reduced land depreciation and transaction taxes. For example, purchasing agricultural land may be extremely lucrative due to its reputation for productivity.

Still, if the buyer declares that they plan to hold it as agricultural land for the next five years, the buyer is excluded from agricultural taxes. Discounts on animal feed, computer and equipment rentals, and veterinarian fees are also included.

Milk, as a fundamental product, is critical to the success of every dairy farm. Indeed, goat milk is consumed by a greater number of people worldwide than cow milk. When combined with proper management and sound decision-making abilities, it will truly be profitable dairy goat farming.

IMPROVING YOUR BUSINESS

Keeping Records on your Goat Farm

Maintaining accurate records is critical to the success of a goat farm. Not only do records assist you in maintaining control over manufacturing, feeding, and profitability, they also ensure that you are complying with food safety and traceability regulations. Most significantly, they're an excellent tool for enhancing your herd and farm's continuous improvement.

Although record-keeping varies by farm, it does not have to be a time-consuming process. It's relatively simple to get started, and the rewards can quickly outweigh the costs.

Maintaining farm records enables us to organize data in almost any way we want. We can quickly produce lists of marketed kids (trace goats in and out), average group weights, average ADG (average daily gain), and average days to market. Weighing each kid three times and measuring gains enables us to make more accurate forecasts of when kids are ready to enter the market.

Additionally, we have our own evidence to demonstrate the efficacy of a modern barn procedure. Lists of kids can be sorted by the dam to create doe production records for each doe, allowing for the calculation of average days to market, total meat generated, or the total value of kids produced. We can fairly accurately determine which goats are earning their keep.

Hoof care is unquestionably one management aspect that has tightened up significantly since we began tracking it. We know that the majority of goats in our herd need hoof trimming approximately every five months and that each doe is inspected during the kidding season.

Does with sore feet spend less time at the feeder and provide less milk for their young, making this a vital task to maintain.

We want to retain replacements from these stalwarts does, but we also need to track them more closely for body condition, hoof health, pregnancy toxemia, and udder condition.

The number of kids raised each year is our primary indicator of doe production, and when this number starts to decrease, does can be sold for meat while still in good body condition.

We assume that the most valuable knowledge is that which is scribbled. Using various colors to highlight certain notes enables us to identify patterns within the herd.

Farm records are essential for establishing the herd's pedigree, implementing the breeding program for herd enhancement, tracking various farm performance metrics, feeding animals economically, culling animals that are not profitable, stocking and selling goods, and computing financial data.

Examples of important records are as follows:
- Ewe History
- Ewe Record
- Ram Record
- Wool Production Record
- Health Record
- Roll Call

Conclusion

There are many people out there that are looking to start up a small business on their own land or become more self-sufficient. Equally, there is something satisfying about building something with your own two hands. You may have dreams of starting your own small herd and creating goods from their gifts, such as milk or fiber. If you have never done this before, it may be difficult to navigate how to get started, where to purchase your seed stock, and what animals to choose.

That's why I have put this book together: to help you determine where to get started and why. Besides details about the more popular breeds of goats today, this book addressed detailed questions you might ask, information that should be given to you, and how to prepare your property to house your new investment.

Goats are adventurous, and it can be difficult to keep them fenced in because they are such proficient escape artists. The key is to keep them happy on their side of the fence and provide them with a proper and safe environment filled with tasty weeds and shrubbery overgrowth, since those are their favorite snacks. If you are new to caring for goats, we have given you insight into how to feed them so that they stay healthy and continue producing goods for you.

Dairy goats are probably the most popular herd today, and there is so much more than just getting goat's milk involved with your new venture. This book has given you a small sampling of recipes that you can create using your new homegrown supply, and there are so many more choices out there! We hope that you will enjoy making cheese, ice cream, and more!

If you choose to start a meat goat herd, you now have a list of some of the best breeds to use to solidify your business. It can be difficult getting started, but once you are in the eye of a recurring buyer, you may receive a contract if you continue to improve your herd's confirmation and your reputation for raising the very best animals.

Another choice, of course, is raising fiber goats. With them, you will enter the world of textiles, and right now, there is more demand than there is production. You received tips on shearing and terminology associated with the industry that will make you sound like a pro.

It can be challenging and a bit scary taking that leap and getting started in your chosen goat business. With determination and quality, you will find your niche in the market. We have explained how important it is to keep records of your venture. This way, you will know how much you are spending and how you are improving the bloodlines of your herd.

Lastly, you now have in dept information about keeping your herd healthy and some challenges that you may face. We never want to, but it helps to be prepared for any emergency! Along with those disease descriptions, this book has included a basic first aid kit supply list that you will want to assemble and keep handy.

Raising your own herd will help you become more self-sufficient and teach you about living off the land. They will give you milk to drink, food to eat and provide you with entertainment and companionship. They can be instrumental in clearing land of unwanted vegetation; they even clear poison ivy!

References

Alphafoodie, S. (2020, July 29). How to make goat cheese (plus FAQs and tips). Alphafoodie. https://www.alphafoodie.com/how-to-make-goat-cheese/

Analida. (2020, January 7). How to make goat cheese recipe - Chèvre. Analida's Ethnic Spoon. https://ethnicspoon.com/how-to-make-goat-cheese-recipe-chevre/

Bradshaw, A. (2015, March 5). Goats for sale - 6 mistakes to avoid when buying goats. Common Sense Home. https://commonsensehome.com/goats-for-sale/

Carroll, R. (2018). Home cheese making: recipes for 75 homemade cheeses. Storey Books. (Original work published 1982)

Castration | Agricultural Research. (n.d.). www.luresext.edu. Retrieved June 5, 2021, from http://www.luresext.edu/?q=content/castration#:~:text=Three%20common%20ways%20to%20castrate

Goats. 2019. Goat Reproduction. 14th August 2019. Goats. Online. https://goats.extension.org/goat-reproduction/

Goats. 2019. Newborn Goat Kids. 14th August 2019. Goats. Online. https://goats.extension.org/newborn-goat-kids/

  Brady, Boyd. 2019. Reproductive Management - Dairy Goats and Sheep. Extension Alabama A&M & Auborn Universities. 12th August 2019. Online. https://www.aces.edu/blog/topics/sheep-goats/reproductive-management-dairy-goats-and-sheep/

Hoschton Animal Hospital. 2016. Signs of Illness in Goats. Hoschton Animal Hospital. Jan 15th 2016. Online. https://www.hoschtonanimalhospital.com/2016/01/15/hoschton-ga-vet-illness-goats

Manitoba Goat Association, Goats and their nutrition. Online. https://www.gov.mb.ca/agriculture/livestock/goat/pubs/goats-and-their-nutrition.pdf

Nicoletti, Paul. 2013. Brucellosis in Goats. MSD Manual Veterinary Manual. Aug. 2013. Online. https://www.msdvetmanual.com/reproductive-system/brucellosis-in-large-animals/brucellosis-in-goats

OIE. 2021. Bovine Tuberculosis. OIE World Organisation for Animal Health. Online. https://www.oie.int/en/disease/bovine-tuberculosis/

 PennState Extension. n.d. Goat Housing and Facilities. PennState Extension. Online. https://extension.psu.edu/programs/courses/meat-goat/basic-production/general-overview/goat-housing-and-facilities

Hobby Farms. Tips for providing your goats the shelter they need. Online. https://www.hobbyfarms.com/tips-providing-goats-shelter-they-need/

Oregon States University. Internal parasites in sheep and goats. Online. https://catalog.extension.oregonstate.edu/sites/catalog/files/project/pdf/em9055.pdf

 Pezzanite, Lynn et al. Common Diseases and Health Problems in Sheep and Goats. Animal Sciences. Online. https://www.extension.purdue.edu/extmedia/AS/AS-595-commonDiseases.pdf

Purdue University Extension. Common diseases in sheep and goats. Online https://www.extension.purdue.edu/extmedia/as/as-595-commondiseases.pdf

Thoen, Charles O. 2014. Tuberculosis in Sheep and Goats. MSD Manual Veterinary Manual. Online. https://www.msdvetmanual.com/generalized-conditions/tuberculosis-and-other-mycobacterial-infections/tuberculosis-in-sheep-and-goats

Fiber Goats – The american goat federation. (n.d.). American goat federation. Retrieved June 8, 2021, from https://americangoatfederation.org/breeds-of-goats-2/fiber-goats/#:~:text=The%20Fiber%20Goat%20industry%20has

Garman, J. (2020, January 4). How long do goats live? - Backyard Goats. Backyard Goats. https://backyardgoats.iamcountryside.com/health/how-long-do-goats-live/

Griffith, K., Rask, G., Peel, K., Levalley, S., & Johnson, C. (n.d.). Raising and showing meat goats! A youth manual for meat goat projects in Colorado! Retrieved June 8, 2021, from https://www.meadowlark.k-state.edu/docs/4h/resources/Meat_Goat_Manual.pdf

Harlow, I. (2015, February 13). Goats for land management. Farm and Dairy. https://www.farmanddairy.com/top-stories/goats-land-management/240237.html#:~:text=If%20you%20want%20to%20clear

Helmer, J. (2020, December 7). Tips for providing your goats the shelter they need. Hobby Farms. https://www.hobbyfarms.com/tips-providing-goats-shelter-they-need/

Kopf, A. : K. (2021, January 19). Goat lice: are your goats lousy? Backyard Goats. https://backyardgoats.iamcountryside.com/health/goat-lice-are-your-goats-lousy/

Manteuffel, R. (2019, August 13). Are goats the new weed whackers? Plenty of people want them to be. The Washington Post. https://www.washingtonpost.com/lifestyle/magazine/using-goats to-clear-land-is-way-more-labor-intensive-than-anyone-can-imagine/2019/08/13/acc17608-b78e-11e9-b3b4-2bb69e8c4e39_story.html

Manuel, A. (n.d.). How to make goat milk candles. Home Guides | SF Gate. Retrieved June 6, 2021, from https://homeguides.sfgate.com/make-goat-milk-candles-74284.html

Metzger, M. (2018, December 14). Winter management tips for goats. sheep & goats. https://www.canr.msu.edu/news/winter-management-tips-for-goats#:~:text=Goats%20that%20are%20properly%20cared

Nutritional feeding management of meat goats | NC State Extension Publications. (2015). Ncsu.edu. https://content.ces.ncsu.edu/nutritional-feeding-management-of-meat-goats

Nuwer, R. (2014, March 26). Never underestimate a goat; it's not as stupid as it looks. Smithsonian; Smithsonian.com. https://www.smithsonianmag.com/science-nature/never-underestimate-goat-not-stupid-looks-180950265/

Pesaturo, J. (2014, February 10). Goat milk ice cream bases |. Our One Acre Farm. https://ouroneacrefarm.com/2014/02/10/goat-milk-ice-cream-bases/

Ploetz, K. (2013, July 16). The law | rspca.org.uk. Rspca.org.uk. https://www.rspca.org.uk/adviceandwelfare/farm/farmanimals/goats/law

Poindexter, J. (2017, March 22). Goat fencing: 6 important tips to consider to build the perfect fence. MorningChores. https://morningchores.com/goat-fencing/

Roy's Farm. (2021a, May 17). Raising goats as pets: beginner's guide for raising pet goats. Roy's Farm. https://www.roysfarm.com/raising-goats-as-pets/

Roy's Farm. (2021b, May 17). Why do bucks smell so bad: the secrets of the goaty smell. Roy's Farm. https://www.roysfarm.com/why-do-bucks-smell-so-bad/ Sartell, J. (2018). Feeding baking soda to your goats. Mannapro.com. https://www.mannapro.com/homestead/feeding-baking-soda-to-your-goats

smallholderhollow. (n.d.). Dairy goats: full size vs mini goats. Retrieved May 27, 2021, from http://smallholderhollow.com/dairy-goats-full-size-vs-mini-goats/

Smith, C. (n.d.-a). Creating a first aid kit for goats. Dummies. Retrieved June 8, 2021, from https://www.dummies.com/home-garden/hobby-farming/raising-goats/creating-a-first-aid-kit-for-goats/

Smith, C. (n.d.-b). How and when to shear your goats. Dummies. Retrieved June 8, 2021, from https://www.dummies.com/home-garden/hobby-farming/raising-goats/how-and-when-to-shear-your-goats/

Smith, C. (n.d.-c). What is normal goat behavior? Dummies. Retrieved May 31, 2021, from https://www.dummies.com/home-garden/hobby-farming/raising-goats/what-is-normal-goat-behavior/

The Hay Manager. (2018, October 31). Keeping goats warm in the winter. The Hay Manager. https://www.thehaymanager.com/goat-and-sheep-round-bale-hay-feeders/keeping-goats-warm-in-the-winter/

The Law | rspca.org.uk. (2017). Rspca.org.uk. https://www.rspca.org.uk/adviceandwelfare/farm/farmanimals/goats/law

Tilley, N. (n.d.). StackPath. Www.gardeningknowhow.com. Retrieved June 4, 2021, from https://www.gardeningknowhow.com/composting/manures/goat-manure-fertilizer.htm#:~:text=Goat%20manure%20is%20virtually%20odorless

Treehugger. (n.d.). Learn how to feed and tend goats on the small farm. Treehugger. Retrieved June 1, 2021, from https://www.treehugger.com/feed-and-tend-goats-3016793#:~:text=Hay%20is%20the%20main%20source

Vet, S. C. (2019, February 11). Three diseases all goats owners should be aware of, test for, and work to prevent. Sale Creek. https://salecreek.vet/three-diseases-all-goats-owners-should-be-aware-of-test-for-and-work-to-prevent/

Wetherbee, K. (n.d.). Raising dairy goats and the benefits of goat milk. Mother Earth News. Retrieved June 2, 2021, from https://www.motherearthnews.com/homesteading-and-livestock/benefits-of-goat-milk-zmaz02jjzgoe

Wikipedia Contributors. (2019a, February 27). Cashmere goat. Wikipedia; Wikimedia Foundation. https://en.wikipedia.org/wiki/Cashmere_goat

Wikipedia Contributors. (2019b, October 27). List of goat milk cheeses. Wikipedia; Wikimedia Foundation. https://en.wikipedia.org/wiki/List_of_goat_milk_cheeses

Wolford, D. (2014a, February 16). Homemade udder & teat wipes. Weed 'Em & Reap. https://www.weedemandreap.com/homemade-udder-teat-wipes-milking/

Wolford, D. (2014b, April 5). How to milk a goat: step by step pictures. Weed 'Em & Reap. https://www.weedemandreap.com/how-to-milk-a-goat/

www.ingramcontent.com/pod-product-compliance
Lightning Source LLC
Chambersburg PA
CBHW030533210326
41597CB00014B/1126